WEATHER MAGIC IN MEDIEVAL SCANDINAVIA

Weather Magic in Medieval Scandinavia explores how people in medieval Scandinavia understood and attempted to influence the weather through magic. Covering the period from c. 800 to 1555 AD, the book reveals how storms, winds, and divine tempests shaped not only medieval imagination but also law, ritual, and power across the North.

Drawing on Old Norse sagas, poetry, law codes, religious texts, archaeology, and early modern trial records, this is the first comprehensive study of weather magic across the medieval Scandinavian world. The book argues that weather magic was not a marginal superstition but a key part of how medieval societies conceptualised human and divine agency within nature. It traces how figures such as Óðinn, *vǫlur*, and saints appear as mediators between people and the elements, and how ideas of magic and miracle blurred in Christian and non-Christian contexts. By combining literary, historical, and material evidence, *Weather Magic in Medieval Scandinavia* illuminates how weather manipulation was bound to performance, voice, and social power, linking magic, seafaring, warfare, and the legitimisation of elite authority. The study offers new interdisciplinary insights into belief, ecology, and cultural exchange in the medieval North.

The book appeals to scholars and students of Old Norse studies, medieval history, folklore, religious studies, and environmental humanities, as well as general readers interested in the intersections of magic, mythology, and climate in the North.

Jennifer Hemphill is an independent researcher in Scandinavian Studies whose work explores the cultural history of magic in Northern Europe. She holds a PhD in Scandinavian Studies from the University of Aberdeen and is a co-founder of the international conference series *Performing Magic in the Pre-Modern North*. Based in Copenhagen, her research combines literary, historical, and folkloric sources to examine how magic shaped medieval and early modern worldviews across the Nordic region.

Routledge Studies in the History of Witchcraft, Demonology and Magic

The Construction of Witchcraft in Early Modern Denmark, 1536–1617
Louise Nyholm Kallestrup

Prophets and Witches
Witchcraft, Gender and Politics in Revolutionary England
Debra Parish

Agents of the Invisible World
Children and Witchcraft in England and New England, 1589–1692
Molly Northcott

The Science of Demons, Second Edition
Early Modern Authors Facing Witchcraft and the Devil
Edited by Jan Machielsen

English Demonologies
The Devil and Witchcraft in Elizabethan and Stuart Thought
Darren Oldridge

Weather Magic in Medieval Scandinavia
Old Norse-Icelandic Traditions in Broader Scandinavian Contexts
Jennifer Hemphill

For more information about this series, please visit: https://www.routledge.com/Routledge-Studies-in-the-History-of-Witchcraft-Demonology-and-Magic/book-series/RSHWDM

WEATHER MAGIC IN MEDIEVAL SCANDINAVIA

Old Norse-Icelandic Traditions in Broader Scandinavian Contexts

Jennifer Hemphill

Routledge
Taylor & Francis Group
LONDON AND NEW YORK

Designed cover image: Woodcut showing a witch creating a storm at sea from Olaus Magnus' 1555 book *Historia de Gentibus Septentrionabilus*. Source: Alamy/Pictorial Press Ltd.

First published 2027
by Routledge
4 Park Square, Milton Park, Abingdon, Oxon OX14 4RN

and by Routledge
605 Third Avenue, New York, NY 10158

Routledge is an imprint of the Taylor & Francis Group, an informa business

British Library Cataloguing-in-Publication Data
A catalogue record for this book is available from the British Library

ISBN: 978-1-041-25302-0 (hbk)
ISBN: 978-1-041-25297-9 (pbk)
ISBN: 978-1-003-74605-8 (ebk)

DOI: 10.4324/9781003746058

Typeset in Sabon
by SPi Technologies India Pvt Ltd (Straive)

CONTENTS

FIGURES

ACKNOWLEDGEMENTS

This monograph stems from the PhD thesis completed during my doctoral studies at the Centre for Scandinavian Studies at the University of Aberdeen, which I submitted and successfully defended in the latter half of 2024. My time at the Centre was an absolute privilege, and I remain deeply grateful for the years I spent living and studying both in Aberdeen and in Scandinavia.

First and foremost, I extend my deepest gratitude to my two wonderful supervisors. Hannah, thank you for your unwavering support from the very beginning and for embracing such a fascinating and joyful topic of research. Your patience, enthusiasm, encouragement, and expertise have been invaluable, and the friendship we developed throughout my PhD has been a cherished part of the journey. Your guidance has strengthened my writing in ways I am immensely proud of. Ralph, thank you for your rigorous feedback and for always challenging me to think more critically. Your engagement has made me a much stronger academic, and I have greatly enjoyed our in-depth conversations on magic and the Old Norse world. I genuinely miss our supervision sessions.

This project was also shaped by a year-long research visit to Uppsala University and Stockholm University. This visit was made possible through financial support from the University of Aberdeen's DHP postgraduate funding, the University of Aberdeen Erasmus+ programme, and the Viking Society for Northern Research. Although the outbreak of COVID-19 drastically altered my plans, the resources I accessed at both university libraries were invaluable, and the relationships I developed with faculty, staff, PhD students, and postdocs significantly enriched my research experience.

I am grateful to the Centre for Scandinavian Studies for fostering such an inclusive and supportive environment. To my friends and colleagues from around the world, thank you for your encouragement and companionship throughout this journey. Your support, both academic and personal, made an enormous difference. I am especially thankful to my wonderful group of Norwegian friends, Solveig, Heidi, Ingrid, and Hans, for the many hours of stimulating conversation and uplifting breaks from the thesis.

Through the University, I also had the pleasure of meeting Ségdae and Selu who, together with Solveig, joined me in creating the Performing Magic in the Pre-Modern North conference series and in publishing our first edited volume, *Performing Magic in the Pre-Modern North: Practice and Transgressions* (Hemphill, Richardson-Read, and Wang 2024). Their generosity, creativity, and shared enthusiasm for the subject made this collaborative work one of the great joys of my doctoral years.

I would also like to thank Melanie Viney, whom I likewise met during my time in Aberdeen. Melanie provided exactly what every long PhD deserves and rarely receives: well-timed humour, much-needed bakes, and an unwavering commitment to delighting in enchanted worlds. She also encouraged my book-collecting habit with far too much enthusiasm, and for that I remain both grateful and only slightly compromised.

I thank my PhD examiners, Professor Stephen Mitchell and Dr Sam Newington, for their thoughtful engagement with my thesis. I am especially grateful to Professor Mitchell for his incisive comments, some of which have been incorporated into this book.

Beyond the academic community, I am deeply appreciative of my family in the United States. To my dad, Ross, and my stepmother, Pam, thank you for supporting my entire academic adventure, both domestic and international. To my brothers, Robert and Will, and my sister, Stephanie, thank you for always being there, no matter the distance. Most importantly, to my mom, Barbara, thank you for encouraging my dreams of living in Europe, for supporting my academic pursuits, and for continually urging me toward the finish line. I am profoundly grateful to my husband, Scott, whose love, support, and encouragement sustained me throughout this entire process. Balancing the demands of the PhD alongside motherhood and full-time work was only possible because of you. I love you to the moon and back.

This monograph is dedicated to my two daughters. I hope you always follow your dreams, no matter how impossible they may seem.

ABBREVIATIONS

The abbreviation of the titles of Old Norse sources follows the abbreviations that are used in the online Dictionary of Old Norse Prose and are listed below. Abbreviations for the primary sources that are not featured in the Dictionary of Old Norse prose have been selected for the present monograph. Additionally, abbreviations for the titles of secondary source collections are included. The full titles of all primary sources referenced in this monograph can be found listed in the bibliography, under 'Primary Sources'.

Bósa	*Bósa saga ok Herrauðs konungs*
BS	*Byskupa Sögur*
CSI	*The Complete Sagas of the Icelanders Including 49 Tales*
CV	*Cleasby-Vigfússon's Icelandic-English Dictionary*
Dudo	*De moribus et actis primorum Normanniæ ducum auctore Dudone Sancti Quintini decano*
Eir	*Eiríks saga rauða*
Eb	*Eyrbyggja saga*
Fsk	*Fagrskinna*
Finnb	*Finnboga saga ramma*
Flóam	*Flóamanna saga*
FSN	*Fornaldarsögur Norðrlanda*
Fbr	*Fóstbrœðra saga*
Fær	*Færeyinga saga*
GD	*Gesta Danorum*
Gísl	*Gísla saga Súrssonar*
GBp	*Guðmundar saga dyra*
GHr	*Gǫngu-Hrólfs saga*

Hallfr	*Hallfreðar saga vandræðaskálds*
Harð	*Harðar saga ok Hólmverja*
Hálf	*Hálfs saga ok Hálfsrekka*
Hkr	*Heimskringla*
HGS	*Historia de Gentibus Septentrionalibus*
Hrólf	*Hrólfs saga kraka*
ÍF	*Íslenzk Fornrit*
JMSV	*Jómsvíkinga saga*
Jóns	*Jóns þáttr biskups Halldórssonar*
Ket	*Ketils saga hængs*
Ldn	*Landnámabók*
Laxd	*Laxdœla saga*
Nj	*Njáls saga*
NgL	*Norges gamle love indtil 1387*
Norn	*Norna-Gests þáttr*
ÓH	*Ólafs saga Helga*
ÓT	*Ólafs saga Tryggvasonar*
Orkn	*Orkneyinga saga*
Orv	*Orvar-Odds saga*
Saxo	*Gesta Danorum*
SKALD	*Skaldic Poetry of the Scandinavian Middle Ages*
StSt	*Sturlaugs saga*
Vatn	*Vatnsdœla saga*
Vígl	*Víglundar saga*
Vols	*Volsunga saga*
Þorláks	*Þorláks saga byskups*
ÞorstVik	*Þorsteins saga Víkingssonar*

INTRODUCTION

Over the past two decades, the field of Old Norse studies has witnessed a significant surge in academic research dedicated to the examination of magic. Among the prevalent motifs in Old Norse literature, the magical manipulation of weather stands out as one of the most frequently explored themes. Weather-magic performance finds expression across all genres of Old Norse saga literature. By the thirteenth century, the sagas recurrently depict practitioners of Northern magic as possessing the uncanny ability to conjure abnormal weather conditions, both on land and at sea.

In *Laxdœla saga*, for instance, a family deeply versed in magical arts summons a treacherous storm at sea, while an elderly woman featured in *Gísla saga Súrssonar* marshals her knowledge in magic to trigger an avalanche that lays waste to her adversary's farmhouse. Numerous poetic verses allude to incantations capable of quelling perilous weather patterns, with a notable example found in Óðinn's compendium of spells in *Hávamál*. This extensive catalogue of instances that showcases weather magic within Old Norse literature and poetry presents scholars with an exceptional opportunity to explore the intricacies of this specific form of magic.

The overarching objective of this study is to provide both a broad overview and a detailed analysis of weather magic as a cultural construct in the Scandinavian Middle Ages, with a primary focus on Icelandic-Norse sources situated within their wider Scandinavian context. The temporal scope encompasses the period spanning approximately 800–1555 AD, commencing with the 'Viking Age', traversing the Christianisation of medieval Scandinavia, witnessing the rise of medieval Icelandic saga composition, and culminating

DOI: 10.4324/9781003746058-1

in 1555 with the publication of Olaus Magnus's work, *A Description of the Northern Peoples*, when the Scandinavian perception of magic becomes increasingly aligned with the Continental European concept of 'witchcraft'.

The research underpinning this project is structured around a rich variety of sources, encompassing textual, archaeological, legal, linguistic, and folkloristic materials. This interdisciplinary approach aims to yield a comprehensive understanding of weather magic as it manifested during the Scandinavian Middle Ages. By piecing together the wide range of source material, this project aims to shed new light on the cultural beliefs and practices surrounding weather magic in this historical context.

Historiography

Old Norse Magic

The sources outlined below are not intended to be an exhaustive review of the study of Old Norse magic, but rather an overview of the key studies that are foundational to the present study. Additional works on the scholarship of Old Norse magic are introduced and analysed throughout the book, where relevant. Magic in the Viking and early Middle Ages has long been a topic of fascination for scholars but has 'proved to be an elusive topic for earlier generations, only slowly precipitating out of the evidence as much more than a shadowy survival of Norse heathendom' (Mitchell 2011: 6). However, there has been an increase in academic works focusing on Old Norse magic over the past few decades (Mitchell 2011: 6), with an additional surge of studies being published in the early-to-mid 2000s. The scholarly attention to Old Norse magic has only grown, and beginning in 2020, there have been several books published on the topic[1] and multiple PhD theses focusing on pre-Modern Scandinavian magic, a trend that shows no signs of slowing down.

The most in-depth study of magic as it is presented in Old Norse saga literature comes from François Xavier Dillmann's 1986 thesis (published in 2006), *Les magiciens dans l'Islande ancienne: Études sur la représentation de la magie islandaise et de ses agents dans les sources littéraires norroises*. Through a comprehensive analysis of the saga corpus, Dillmann explores the literary portrayal of medieval Icelandic magic and its practitioners. His study delivers an analysis of all angles of magic as it is presented in the sagas, and considers the literary practitioners of magic as an essential part of the socio-cultural fabric of medieval Iceland. In his thesis, Dillmann suggests that the episodes of magic found in the saga corpus can provide insight into the medieval Icelandic perception of actual pre-Christian magic and the individuals who wielded it, shedding light on the intricate relationship between magical practices and the collective imagination of medieval Iceland.

In 2002, Catharina Raudvere published a chapter on the Old Norse magical term *trolldómr* in a monograph about general magic in the Middle Ages (Raudvere 2002), and then in 2003, she published her extensive book-length study on Old Norse magic, *Kunskap och insikt i norrön tradition: Mytologi, ritualer och trolldomsanklagelser* (Raudvere 2003). Through the lens of a historian of religion, Raudvere takes on the large task of examining episodes of magic in Old Norse saga literature and the terminology used to describe these events, while she cautions against reading these magical episodes as true reflections of actual pre-Christian practices, as is suggested in Dillmann's study. Instead, Raudvere proposes that through a careful reading of the source material, one can unravel some aspects of pre-Christian traditions that influence the magic scenes, as well as the Christian thought that influences the portrayal of magical events in Old Norse texts, rather than interpreting all scenes of magic as accurate reflections of pre-Christian practices.

In addition to the new text-based studies focusing on Old Norse magic produced in the early 2000s, archaeologists began to assess the physical material of the Viking Age that could be linked to magical practices, applying new, interdisciplinary methodologies like cognitive archaeology and queer theory. *The Viking Way: Religion and War in the Late Iron Age Scandinavia* by Neil Price (2002) explores the intersection of religion, warfare, and society in Viking Age Scandinavia, with a particular focus on magic in a Norse and wider 'circumpolar' context. Price presents his interpretation of the Viking Age mind-set, which stems from his interdisciplinary approach to archaeological evidence, historical texts, and comparative studies of other circumpolar cultures. His book explores the religious beliefs and magical practices of Viking Age Scandinavians and their neighbours by analysing the complex Norse mythology, the role of the Viking Age gods (with a particular focus on Óðinn and his magical knowledge), and how religious beliefs shaped the worldview of Viking Age Scandinavians. Price's work was of game-changing significance in the field of Old Norse magic, especially his investigation of the magical practice of *seiðr* (to be defined and discussed in the first chapter of the present monograph that focuses on magic terminology) and its role in Viking Age life. He also explores the potential connection between *seiðr* and the magical practices of the Norse's Indigenous[2] neighbours, the Saami. In 2019, Price released a revised and expanded version of his monograph, with the new title of *The Viking Way: Magic and Mind in Late Iron Age Scandinavia* (Price 2019). The conscious addition of adding the term 'magic' into the revised book's new title reflects the growing attention to the study of Old Norse magic. Although most of the text and arguments remain identical to the first edition, the second edition includes a final chapter that retrospectively reviews the notions of magic that were presented in the original publication, where Price himself admits that 20 years later he does not necessarily agree with the study's original framework.

Following Price, Brit Solli's archaeological study, *Seid: Myter, sjamanisme og kjønn i vikingenes tid*, published in 2002, offers a novel analysis of *seiðr* in the Viking Age by applying Queer Theory. This new approach to research on Viking-Age practices allows Solli to explore the complex intersections of gender, sexuality, and magic within Old Norse religious practices. By examining relevant archaeological materials, Solli assesses how gender roles and sexual identities were intertwined with the practice of *seiðr*, challenging traditional perceptions of Viking Age society.

Solli's work is particularly significant as it is the first to employ Queer Theory in this context, providing a new methodology through which to understand the fluidity and non-binary aspects of gender and sexuality in Old Norse culture. Her analysis reveals that the practitioners of *seiðr*, who are often portrayed in the Old Norse sagas as transgressing normative gender boundaries, played crucial roles in their communities. Solli demonstrates that this transgression is not only a literary motif but is also supported by archaeological findings, such as grave goods and burial practices that indicate a complex and inclusive view of gender.

Solli's study thus not only deepens our understanding of *seiðr* but also contributes to broader discussions on the roles of gender and sexuality in historical societies. By highlighting the cultural significance of these fluid identities, Solli's research challenges conventional historical narratives and opens new avenues for exploring how medieval societies understood and integrated diverse gender and sexual identities into their social and religious frameworks.

Possibly the most important contribution to the study of Old Norse magic published in the past decade, especially to the present study, is Stephen Mitchell's *Witchcraft and Magic in the Nordic Middle Ages* (2011). Mitchell's approach diverges significantly from those of Dillmann, Price, and Solli by offering a more comprehensive analysis that integrates historical, cultural, and social dimensions of magic. Unlike Dillmann, whose work primarily focuses on the literary portrayal and socio-cultural role of magic practitioners within the Old Norse saga corpus, Mitchell expands the scope to include a broader spectrum of magical practices and their societal implications across various sources. Price's interdisciplinary methodology, which intertwines archaeology, history, and comparative studies of circumpolar cultures, concentrates heavily on the Viking Age and the role of *seiðr*, particularly in warfare and gender contexts. In contrast, Mitchell's work extends beyond these confines, providing a thorough historical overview that contextualises magic within the broader medieval Scandinavian society and its evolution over time. Mitchell distinguishes his work by diving into the relationship between magic, religion, and science, offering a detailed exploration of the scholarly debates surrounding these concepts. His methodology includes a critical analysis of key terms and concepts in an Old Norse context, which clarifies the distinctions and overlaps between different forms of magic.

One of Mitchell's notable contributions is his examination of weather magic, a section that, although briefly covered, is significantly important to the present study (Mitchell 2011: 65–66). He discusses the practical applications of weather magic in Scandinavian daily life, such as its impact on agriculture, livestock, fishing, and sailing, all of which were critical aspects of survival in medieval Scandinavia. This focus on the everyday utility of weather magic, reflecting its prevalence in Old Norse texts, provides a fresh perspective compared to the more mythologically and ritually centred analyses of his predecessors. Thus, Mitchell's monograph provides a broader, more integrated approach that highlights the pervasive and practical nature of magic in medieval Scandinavian society, setting a new standard for future research in the field.

Old Norse Weather Magic

Although weather magic is one of the most commonly employed magical motifs in Old Norse sources and it appears as a minor element of many studies on the general study of Old Norse magic, weather magic on its own is a surprisingly understudied topic. Most of the research on weather magic conducted prior to this study has focused on the god Þórr or only on the weather magic that is represented in the *Íslendingasǫgur*. Richard Perkins' monograph *Thor the Wind-Raiser and the Eyrarland Image* (2001) is one of the first publications that is completely dedicated to the study of weather magic in Viking Age and medieval Scandinavia. In his book, Perkins delivers a brief overview of Norse wind magic and wind gods, and then turns his focus completely to his argument that Þórr was understood to have the ability to manipulate the winds and that there is surviving physical evidence of this belief. Perkins' study is influential to the present study in that it employs a similar interdisciplinary approach to the examination of Scandinavian weather magic. In his analysis, Perkins incorporates a wide variety of source material, including episodes from Old Norse saga literature, poetry, folklore, and archaeological materials. In the following chapters of this monograph, I will engage with Perkins' representation of Þórr as a wind god and his proposal that there are three surviving figurines of Þórr from the Viking Age that reflect a belief in his ability to manipulate the weather. However, there are already many robust studies published in the last two decades that focus on Þórr's relationship with the weather (for example, Perkins 2001 and Taggart 2017) and there are no medieval examples where the god's association with the weather is attributed to his explicit knowledge in magic. I therefore chose to focus more extensively on the Old Norse deity Óðinn, rather than Þórr, and his reputation for using weather magic, since the surviving primary source material explicitly attributes his ability to manipulate the weather to his knowledge in magic.

A few years following Perkins' publication on Þórr as a wind-raising god, Astrid E. J. Ogilvie and Gísli Pálsson collaborated on and published an article, 'Mood, Magic, and Metaphor: Allusions to Weather and Climate in the Sagas of Icelanders' (2003), which analyses episodes of weather magic and general episodes of weather in the *Íslendingasǫgur*. The authors suggest that weather is used in these texts as metaphors and as a literary device to establish the mood of the saga. They further extend their argument by suggesting that the episodes of general weather and weather magic in this group of sagas 'must also reflect the ways in which medieval Icelanders thought and talked about the environment and their place within it' (Ogilvie and Pálsson 2003: 269), thus connecting the natural landscape and cultural landscape in medieval Iceland. Although a very useful study on the employment of weather magic as a literary motif in medieval Scandinavia, Ogilvie and Gísli Pálsson's article solely focuses on episodes of weather magic found in the *Íslendingasǫgur* and does not provide a deep analysis of the literary context surrounding these magical performances, such as the performers, how the act is performed and terminology, which will all be addressed in the present study.

The analysis of weather magic in the *Íslendingasǫgur* continues with the publications of Terry Gunnell's article '"Magical Mooning" and the "Goatskin Twirl": "Other" Kinds of Female Magical Practices in Early Iceland' (2014) and Bernadine McCreesh's chapter on weather magic in her book *The Weather in the Icelandic Sagas: The Enemy Without* (2018) that surveys the overall inclusion of weather as a motif in the *Íslendingasǫgur*. In his article on female magic performance, Gunnell (2014) investigates the representations of specific weather-magic performances in the *Íslendingasǫgur*, which he coins 'magical mooning' and 'the goatskin twirl'.[3] Gunnell concludes that although it is difficult to draw concrete conclusions about the performance of weather magic as it is represented in these family sagas, both 'magical mooning' and the 'goatskin twirl' appear to reflect practices of non-Christian weather magic to some extent. Both of these weather-magic performance types and Gunnell's arguments will be engaged with in depth in the following chapter on the performance of weather magic (Chapter 3 of this monograph).

Bernadine McCreesh's (2018) abovementioned monograph that focuses on the employment of the weather as a motif in the *Íslendingasǫgur* includes a very useful chapter that offers an overview of every episode of weather magic that is featured in the *Íslendingasǫgur*. Although the scope of her project does not allow for a very deep analysis of each episode of weather magic and is limited to this specific group of sagas, her list of weather-magic episodes and her initial arguments on the topic are invaluable to the present study. McCreesh's monograph also outlines the portrayal of weather miracles in the *Íslendingasǫgur*, all of which will be engaged with in the following chapter on weather miracles in comparison to weather magic (Chapter 5).

Research on the connection between the Saami and their special knowledge of weather magic has been very popular throughout the past two decades, although there have not been many full studies specifically dedicated to the topic. In the monograph *Raising the Wind: The Legend of Lapland and Finland Wizards in Literature*, Ernest J. Moyne (1981) explores this widespread connection between the Saami and weather magic throughout the ages, although his section on medieval magic is only allocated a few pages. There are many other studies that touch on the medieval portrayal of the Saami's reputation for being skilled at weather magic, for example, Solveig Marie Wang's (2023) recent monograph *Decolonising Medieval Fennoscandia: An Interdisciplinary Study of Norse–Saami Relations in the Medieval Period* or Ellen Alm and Rune Blix Hagen's (2020) chapter 'Sámi Magic and Rituals from *Historia Norwegie* to Johannes Schefferus, *c.* 1150–1680', many of which will be examined in the present book's chapter that focuses on Saami weather magic (Chapter 4).

Methodology

To my knowledge, this monograph is the first research project dedicated to the study of weather magic as it is represented across all the surviving source material from medieval Scandinavia. To examine medieval Scandinavian weather magic, I employ a multi- faceted approach that integrates various types of sources, including Old Norse saga literature, Old Norse poetry, continental texts, early modern witch trial records, church writings and correspondences, folklore, and archaeology. The use of a wide scope of source material ensures a comprehensive exploration of weather magic within the context of medieval Scandinavian society. To develop a full understanding of weather magic in this cultural milieu, I closely analyse instances of weather magic found in these diverse sources. Key contextual details, such as the terminology used, practitioners involved, methods of magic employed, the location of the magical events, the targets of the weather magic, and the motivations behind it, are carefully noted and scrutinised. By employing this methodology of examining interdisciplinary sources, I aim to provide a nuanced examination of weather magic within medieval Scandinavian literature, shedding light on its cultural and historical significance in the broader context of the era.

Primary Sources

The texts that I focus on in my analysis of weather magic include Old Norse saga literature, both the Poetic and Prose Eddas, a small sample of skaldic poetry, Old Norse legal texts, a selection of Latin texts from medieval Scandinavia, including Saxo Grammaticus' *Gesta Danorum*, and a selection of relevant sources from continental Europe, such as Adam of Bremen's *Gesta Hammaburgensis ecclesiae pontificum*. Although this book is interdisciplinary in range and examines a wide scope of written texts and a selection of archaeological materials, it heavily focuses on the portrayals of weather magic in Old Norse saga literature, as this collection of literature holds the most abundant examples of weather manipulation. Due to the scope of weather-magic motifs in Scandinavian literature and the size of the Old Norse corpus, I only focus on scenes that are set in the Scandinavian world, outside of a few select scenes set in Europe, which provide supporting context to the analysis of weather-magic scenes set in Scandinavia.

To employ a comprehensive overview of all the contemporary source material available in my analysis of weather magic in the Scandinavian Middle Ages, relevant archaeological materials, such as runic inscriptions, will also be discussed, although possible physical evidence of weather magic in medieval Scandinavia is sparse. The limited archaeological material that is analysed will be introduced in the chapters in which they appear.

The inclusion of multiple types of source material enables a comprehensive analysis of not only episodes directly related to weather magic but also broader references to magic within these diverse literary traditions. By employing a wide-ranging review of the interdisciplinary materials available, this research endeavours to shed new light on the role and significance of weather magic within medieval Scandinavian society, offering new insights into the cultural, social, and religious aspects of magic in the region.

Old Norse Poetry

Eddic poetry is the most important sub-genre of Old Norse poetry for the study of medieval Norse magic. Almost all of the poems that make up the collection known as the Poetic Edda survive in the Codex Regius (GKS 2365 4°), an Icelandic manuscript that was created around the 1270s. Although the composition of these poems is hard to date, the poetry's linguistic elements suggest that the poems do not predate the Viking Age (pre-800) and were most likely formed between the mid-ninth and mid-thirteenth centuries (McKinnell 2005: 37). The magic represented in the Eddic poems is multifaceted and is portrayed as an integral part of Norse mythology.[4] Divine magic is heavily featured throughout the corpus, with a particular emphasis on Óðinn's knowledge of magical spells. For example, the Eddic poem *Hávamál*, which will be discussed in detail in the following chapters, concludes with a list of spells that the god has mastered and includes his ability to calm the seas. Rune magic is also a common feature in this poetic corpus, often appearing as a magical means for divination, protection, and magical power. In the poem *Sigrdrífumál*, where the Valkyrie Sigrdrífa provides advice to the legendary hero Sigurðr, Sigrdrífa outlines a long list of magical runes that Sigurðr needs to know, alongside details of their intended use, which includes a rune that will calm the weather at sea.

Skaldic poetry is a highly intricate and stylised form of poetry that was composed and recited by skalds during the Viking Age and medieval period in Scandinavia. This type of poetry has a complex structure and employs the extensive use of kennings (a compound expression with metaphorical meaning). While not a dominant theme by any means, skaldic poetry occasionally references magical practices. For example, in Sturla Þórðarson's *Hrafnsmál* (composed after c. 1263), the Scottish King Alexander is portrayed as performing weather magic to trap a Norwegian fleet in the Firth of Clyde (SKALD 2: 740–741). Although the lack of magical references in skaldic poetry makes it difficult to draw any concrete conclusions about magic in medieval Scandinavia, the present study will analyse kennings that incorporate weather terminology to examine Óðinn's possible reputation for being skilled in weather magic.

Rímur, which literally translates to 'rhymes' in English, is another form of Old Norse poetry that is particular to Iceland. The earliest collection of *rímur* survives in the Kollsbók: Codex Guelferbytanus 42.7. Augusteus quarto manuscript, dated to 1480–1490. However, because these poems repeat episodes that are portrayed in Old Norse saga literature, including the scenes that depict magical performances, the present study will not focus on the episodes of weather magic illustrated in *rímur* poetry.

Old Norse Saga Literature

Neil Price notes that the corpus of Old Norse sagas is 'saturated with references to [magic]' (Price 2002: 69), and many of these literary works that feature weather magic will be the focus of this book. Several of these sagas are rooted in a historical and realistic setting, and reflect the social and cultural norms of medieval Iceland. The magic that is portrayed in the Old Norse saga corpus is mostly depicted as part of everyday life in medieval Scandinavia, rather than a fantastical or extraordinary occurrence. In these narratives, episodes of magic are integrated into the fabric of the society and are seen as a natural part of the characters' world. Having a basic understanding of the Old Norse saga tradition is essential in understanding the cultural importance of weather magic in medieval Scandinavia. As I am hoping that the present study will also be of interest to readers outside of the Old Norse academic community, a brief introduction to these narratives, how they are generally classified, and the use of magic as literary motifs in each grouping should prove useful.

In modern scholarship, the Old Norse sagas are generally divided into a taxonomy focusing on the text's subject matter. The *konungasǫgur*, also called the kings' sagas in English, were composed in the twelfth through fourteenth centuries in Iceland and Norway, and their events mainly focus on the lineage of Scandinavian kings. The magic portrayed in the *konungasǫgur* is often used in connection to specific historical events and battles. The episodes of magic that are depicted in this subgroup mainly relate to political necessity, which demonstrates how magic could be used as a political tool. *Konungasaga* narrators used magic as a tool that is performed by leaders and political figures, and weather magic is often featured as an instrument employed by non-Christians to resist the inevitable conversion and political rule of the Norwegians. The emphasis on the magical knowledge of political leaders sets this group apart, as the remainder of the Old Norse sagas tend to focus more on the magic of individual characters who have more personal, individual motives like revenge, rather than large-scale political prowess. This group also includes less personal transformation via magic and divination, whereas the main characters of other saga genres tend to undergo personal transformation or have their fates influenced by magic. Instead, in the kings' sagas, magic and divination determine the fate of royal families or realms and influence political outcomes and whole communities, rather than individual destinies.

The *Íslendingasǫgur*, commonly referred to as either the 'Sagas of the Icelanders' or 'the family sagas' in English, are a group of around 40 sagas in which the events are set in Iceland during the Settlement Period from the 870s until around 1030 (Phelpstead 2020: 7). The magic that is incorporated into these narratives is often performed by individual magic practitioners, appearing in the form of prophetic dreams, curses, or rituals. There is less of an emphasis on the magic of supernatural beings, gods, or other mythical creatures, focusing instead on human characters who possess a special, learned knowledge of magic. The practice of magic is seamlessly integrated into the daily lives of characters, where it appears to be shown as part of the natural world rather than being extraordinary. The Sagas of the Icelanders are riddled with episodes of weather magic, where it is portrayed as part of the everyday lives of the saga characters and integrated into the realistic setting of medieval Scandinavia. The magic practitioners of this grouping of sagas are depicted as having a wide repertoire of skills and knowledge that can alter the weather. Like magic in general, the weather magic that is included in the *Íslendingasǫgur* is mostly performed by human characters, and their knowledge of weather manipulation appears to be an inherent part of the character's identity.

The *fornaldarsǫgur*, also known as 'the legendary sagas' in English, are concerned with the 'ancient' times (i.e. the time before the settlement of Iceland in the 870s) and take place in the Old Norse world and Continental Europe. While other saga groups often strive for a more realistic portrayal of life, the *fornaldarsǫgur* sometimes embrace more mythical and legendary motifs.[5] In these sagas where the line between myth and history is often blurred, magic is employed as an integral part of the storytelling. These narratives are often set in otherworldly settings, like enchanted forests, hidden realms and fantastical landscapes, which often results in the saga characters encountering magical and otherworldly creatures during their legendary adventures. In these texts, magic often serves as a means to bridge the gap between human characters and the supernatural. Heroes often meet supernatural beings in the narrative and receive their otherworldly assistance through the form of guidance and prophecies. Magical objects, such as enchanted swords, rings, and amulets, also often feature in the *fornaldarsǫgur* narratives. These objects often grant the saga heroes superhuman abilities or protect them against harm.

The bishops' sagas, or *biskupasǫgur* in Old Norse, are centred on the lives of Icelandic bishops. The magic in these sagas is very much viewed through the lens of Christian belief and theology, and the events that take place in the saga are very much rooted in a Christian context. Instead of the traditional forms of magic that are featured in the other saga genres, the *biskupasǫgur* mostly feature miracles and divine interventions performed by Christian saints and bishops. Although examples of these miraculous acts have the same results as some of the non-Christian magic featured in other groups, like manipulating the weather, these events are seen as acts of divine power rather than magical practices. The difference between magic and the miraculous will be further explored in Chapter 5. Although this kind of saga is often overlooked in the study of Old Norse magic, the texts that feature the lives of Saint Þorlákr Þórhallsson (*Þorláks saga helga*) and Jón Halldórsson (*Jóns þáttr biskups Halldórssonar*) are essential in the present study, as they provide examples of both weather miracles and weather magic, which will be compared to similar episodes of weather magic that occur in other saga groups.

The final category of saga that is included in the present study is the *riddarasǫgur*, or the chivalric sagas. This group of sagas is heavily influenced by the chivalric ideals and romance literature of medieval continental Europe and is different to the other groups of saga literature, which are mostly rooted in Norse culture and history. The magic that is featured in these texts often takes the form of enchantments, love potions, and magical intervention.

Legal Texts from Medieval Scandinavia

Although no surviving mentions of weather magic are found in the legal texts from medieval Scandinavia, these texts are important to the present study in that they provide context for the legal ramifications of performing magic or seeking out magical services during this time period. In the medieval Icelandic legal text known as *Grágás*, first written down in the thirteenth century, Icelanders are forbidden from worshipping *heiþnar vættir*, or 'heathen beings', with a penalty of lesser outlawry (*Grágás* I: 22). In the same section, the text describes the penalties for using or teaching different severities of magic: 'Ef maþr ferr með galldra eþa gørningar. eþa fiolkýngi. þa ferr hann með fiolkyngi ef hann queðr þat eþa kennir. eþa lætr queða. at ser eþa at fe sinv' (*Grágás* I: 22) [If someone uses spells or [magic or magical knowledge] – he uses magic if he utters or teaches someone else or gets someone else to utter words of magic over himself of his property (Dennis et al. 2006 39)]. The penalty for using this less serious form of magic is lesser outlawry and has to be prosecuted by a panel of twelve. The legal text continues by describing the ramifications for practising a more severe form of magic that results in sickness or death: 'Ef maþr ferr með fordæs skap. þat varþar scogGang. þat ero fe eþa mavnnvm.' (*Grágás* I: 22) [If a man practices [a more serious form of magic], the penalty for that is full outlawry. It is [a more serious form of magic] if through his words or his magic a man brings about the sickness or death of livestock or people. (Dennis et al. 2006: 39)].[6] Magic that resulted in sickness and harm is assigned the penalty of full outlawry, also determined by a panel of twelve. The law code then goes on to ban people from using or producing magical objects, with the penalty of lesser outlawry.

The ban on false accusations of magic can be found in various medieval legal codes from Scandinavia. Similar bans on false accusations are found in the Icelandic *Grágás*, the Norwegian *Gulaþingslög* that was written down in the late thirteenth century and the Danish *Skånske lov* that was written down in the early thirteenth century, where there was an emphasis on the importance of presenting evidence for such claims. This recurring ban aimed at discouraging baseless accusations of practising magic, and shows a recognition of the potential for the misuse of such accusations.

It is important to note that while references to magic do exist in the medieval Scandinavian law codes, they are relatively limited in scope and do not provide detailed regulations for magical practices during the time period. As previously noted in the overview of Old Norse literary sources, magic was more commonly addressed in the Old Norse saga corpus and eddic poetry, where it plays a more prominent role in the narratives and mythology. Instead, Scandinavian legal codes primarily focus on matters related to property, crime, and social order, with magic being a very peripheral concern.

Miscellaneous Medieval Scandinavian Texts

There are several texts written in medieval Scandinavia that do not fit in to the aforementioned sub-categories of literature from Scandinavia, which prove to be essential to the present study. The most studied of these miscellaneous texts is Saxo Grammaticus's *Gesta Danorum*, known as 'The History of the Danes' in English, which is a twelfth-century Latin historical work that presents a detailed account of Danish history and mythology. Stephen Mitchell summarises Saxo's general treatment of magic and magic practitioners as being 'associated with most of the ills of mankind' (Mitchell 2011: 105), where the medieval author uses magic as a connection with 'the lost bygone world of paganism' (Mitchell 2011: 106). Mitchell (2011: 106) notes that Saxo's attention to magic in *Gesta Danorum* is wide-ranging and his narrative includes almost every form of magic that is represented in the Old Norse saga corpus. This is certainly the case in Saxo's inclusion of weather magic. He portrays weather magic when the Bjarmians invoke a magical storm in battle (Saxo i.18.16) and in the same scene, when an old man is able to dispel the Bjarmians' storm with his own magical cloud (Saxo i.18.16).[7] Additionally, the most recent editors of *Gesta Danorum* note that this text is the earliest source where Erik Wind-hat (Ericum uentosi pillei) is referred to by his weather epithet (Saxo ix.4.33: 657), although Saxo includes no mention of his ability to obtain a fair wind from his magical hat.[8]

Konungs skuggsjá, or 'The King's mirror', is an example of speculum literature that was written around 1250. The largest portion of the text is dedicated to the politics and morality on which this type of text traditionally focuses; however, there are several chapters that discuss marvels found in various countries. Structured as a conversation between a father and his son, the text includes many stories about supernatural creatures that are encountered throughout the north. In discussing the marvels of the waters around Greenland, the father tells his son that mermen and mermaid sightings signal that violent storms will follow.[9] The presence of magic amidst the political and moral discourse of *Konungs skuggsjá* reflects the complex perception of magic as a legitimate subject worthy of serious consideration within the cultural context of medieval Scandinavia.

Although it does not include examples of weather magic, the medieval Latin text known as *Historia Norwegiæ* is important to the general study of medieval Norse magic and the present study because of its references to Saami magic. The anonymous author of this text portrays the ability of the Saami to prophesy the future and retrieve lost objects through their magical knowledge. In medieval literature, the Saami are renowned for their ability to control the weather and this text provides further context surrounding their reputation for magical knowledge. By incorporating references to Saami magic within *Historia Norwegiæ*, the author demonstrates a contemporary scholarly acknowledgement of the cultural significance of magic in the medieval Scandinavian world. This inclusion suggests that the author recognised the importance of documenting not only political and social aspects but also magical beliefs and practices prevalent during the time. By incorporating these magical references, the author suggests a belief in the validity of these practices, even if they cannot be explained, which demonstrates an acceptance of the magical aspects of medieval Scandinavia within a historical framework.

In 1555, Archbishop Olaus Magnus, a Swedish writer and cartographer, published *Historia de Gentibus Septentrionalibus* (*A Description of the Northern Peoples*). In this historical and ethnographic work, Magnus writes on the beliefs and practices of people living in Northern Scandinavia. While his text does not focus exclusively on magic, Magnus incorporates many of the magical motifs that are found in Old Norse saga literature, including several episodes depicting weather magic or the magical control over natural elements. For example, Magnus describes a weather-magic practice in Northern Scandinavia in Chapter 16 of Book three (HGS iii.16.173), where he describes the Saami practice of selling magical wind tied into knots of rope; in Chapter 18 of Book three, where he describes Oddi the Dane's ability to 'cause the waves and winds to rage by means of his sorcery' (HGS iii.8.176); and in Chapter 19 of Book three, where Magnus describes the Bjarmian's use of weather magic in attempt to thwart King Ragnar's armies (HGS iii.19.177). Magnus's publication is of particular interest to the present study, in that it is published in 1555, when Scandinavia's attitude towards magic becomes more aligned with the continental approach to witch trials, a change which is represented in the tone of his depictions of magic.

Continental Sources

During the beginning of the period covered in the present study, writings referencing Scandinavia's belief in weather magic began to emerge outside Scandinavia. Additionally, writings on other European weather-magic practices and beliefs outside of Scandinavia were circulating. These sources, originating from various regions in Western Europe, provide invaluable insights into the prevailing attitudes and beliefs surrounding weather manipulation during the Middle Ages. The continental sources below are essential to the present study in that they illuminate the perspectives of Church authorities on weather magic and the contemporary understanding of Scandinavian weather magic circulating in Europe, thus offering a broader contextual framework for the examination of the abovementioned Scandinavian sources.

During the years 815–817, Archbishop Agobard of Lyon wrote a treatise on the local belief in the *tempestarii* – individuals who were believed to be endowed with magical abilities to summon, expel, and manipulate storms. In his writing, Agobard describes that during times of famine or agricultural hardships, local communities would call upon the *tempestarii* to summon good weather to ensure a good crop. Agobard's account illuminates a fascinating discourse surrounding the belief in weather magic during the Middle Ages. Within his treatise, Agobard recounted his encounters with many individuals who earnestly believed that the *tempestarii* were able to control the weather. In response to this prevailing belief, Agobard vehemently asserted the sole authority of God over the weather, denouncing the notion that humans could wield such divine power.

While Agobard's treatise does not offer empirical validation of the existence of *tempestarii* or their practice of weather magic, its very existence suggests a broader cultural acceptance of beliefs in weather magic during the Middle Ages. Agobard's need to articulate and defend his position implies his opinion is a significant divergence from the popular consensus, indicating that the notion of weather magic was not merely relegated to the realm of superstition but held tangible significance within popular belief.

Agobard's writings are significant in understanding the theological stance regarding the existence and belief in weather magic during the Middle Ages. As an esteemed ecclesiastical figure, Agobard's treatise not only serves as a testament to the prevailing attitudes towards weather manipulation around Lyon but also provides invaluable insights into certain theologians' position on such matters. By adamantly refuting the notion of *tempestarii* and attributing weather control solely to God, Agobard rejects the contemporary belief and use of magical practices associated with weather manipulation. His treatise reflects the local doctrinal stance as an opponent of any belief system that challenges the omnipotence of God. Agobard's writings serve as a cornerstone in understanding theologians' efforts to combat perceived heresies and superstitions, and shed light on the intricate dynamics between religious orthodoxy and popular beliefs in medieval Europe.

One of the earliest written records of Scandinavians seeking a change in the weather by appealing to a deity (in this case, Þórr) is found in Dudo of St Quentin's history of the early dukes of Normandy, *De moribus et actis primorum Normanniæ ducum*, which was likely completed in the early eleventh century (Perkins 2001: 20). Dudo recounts a ritual where a group of Norse raiders conduct sacrifices to Þórr, so that they would receive good sailing winds. According to Dudo's account, a divinatory specialist would determine the ritual course of action, and then the sacrificial victims who were chosen by lot were struck on the head with a single blow from an ox-yoke. Subsequently, their brains were examined, and their blood was used to anoint the heads of the men before they set sail, in the hope that this action would appease Þórr and ensure favourable weather:

> Cæterum, in expletione suarum expulsionum atque exituum, sacrificabant olim venerantes Thur, Deum suum. Cui non aliquod pecudum, neque pecorum, nec Liberi Patris, nec Cereris litantes donum, sed sanguinem mactabant humanum, holocaustorum omnium putantes pretiosissimum; eo quod, sacerdote sortilego prædestinante, jugo boum una vice diriter icebantur in capite; collisoque unicuique singulari ictu sorte electo cerebro, sternebatur in tellure, perquirebaturque levorsum fibra cordis, scilicet vena. Cujus exhausto sanguine, ex more suo, sua suorumque capita linientes, librabant celeriter navium carbasa ventis, illosque (v.l. deosque) tali negotio putantes placare, velociter navium insurgebant remis.
>
> *(Dudo 1865: 129–130)*

> Moreover, as the last act on expulsion and departure (?), they used formerly to make sacrifices in veneration of their god Thor. To him they offered neither gifts of sheep nor cattle (?), nor of Father Liber (i.e. 'wine'), nor of Ceres (i.e. 'corn' or 'bread'), but they used to sacrifice the blood of human victims, thinking that to be the most valuable of all offerings. The procedure was this: When a divinatory priest had decided how things should be, the victims were struck viciously on the head with a single blow from an ox-yoke. And then, when the brain of each and every victim, selected by lot, had been crushed by a separate blow, he was spread out on the ground and the fibre (fibra) of the heart, that is to say the vein (vena), was carefully examined on the left-hand side (levorsum). After that, and in accordance with their custom, they would smear their own and their comrades' heads with the drained blood and promptly spread the sails of their ships to the winds, thinking to placate them (i.e. 'the winds'; or, accepting the reading deosque, 'the gods') by such actions. Then they would rapidly ply the oars of their ships.[10]
>
> *(Perkins 2001: 20)*

It is important to note that Dudo himself did not witness these events. However, Perkins (2001: 20) argues that while Dudo was not alive during the events he described, he could draw from the experiences and accounts of the still-living grandchildren of those alive when the events took place and compare these recounted practices to contemporary Scandinavian rituals. Regardless of whether this specific ritual took place exactly as Dudo described it, or if the sacrifice even took place at all, Dudo's writing underscores the fact that the notion that Scandinavians resorted to sacrifices to influence the weather was circulating during his lifetime.

Around a century later, in 1080, Pope Gregory VII wrote a letter to King Harald of Denmark, which demanded that Harald put a stop the practice of attributing inclement weather to his priests:

> Meanwhile there is one matter which has come to our knowledge and which cannot be passed over but must be sternly repressed by apostolic prohibition, namely, that you ascribe to your priests the inclemency of weather, foulness of the air and certain ills of the body ... We therefore direct you by apostolic authority to abolish this pestilent practice absolutely from your kingdom and no longer presume to inflict such disgrace upon priests and clerics.
>
> *(Gregory the Great 1959: 153)*

Moreover, Pope Gregory extends his admonishment to include the accusation of women of having control over the weather, cautioning against any inclination to blame them for inclement weather. The Pope urges King

Harald to exercise mercy and penance rather than harshly punishing innocent women, emphasising the importance of avoiding divine retribution.

> Further, think not that you have a right to proceed by any impious method against women who are condemned by any similar barbarity of heathen practices. Learn rather to avert the divine vengeance by suitable penance than to bring down the wrath of God still more heavily by treating those innocent women with useless severity.
>
> *(Gregory the Great 1959: 154)*

In essence, Pope Gregory's letter to King Harald of Denmark underscores that there was a belief within Scandinavian society in the notion of weather magic, which was thought to be employed in a religious context by religious specialists (Christians in the context of the Pope's letter and non-Christians in the context of Dudo's account) or by human magic practitioners, women specifically. Pope Gregory's letter highlights a shared belief among Scandinavians and broader European communities in the perceived connection between Scandinavians and control of the weather, reflecting the significance of weather magic in Norse culture during the twelfth century.

The *Gesta Hammaburgensis ecclesiae pontificum*, known as the *History of the Archbishops of Hamburg-Bremen* in English, is an important historical treatise written by Adam of Bremen between 1073 and 1076. This text is significant to the present for its early documentation of Norse religious beliefs, particularly those regarding Þórr's perceived control over weather phenomena such as wind, storms, and lightning. Adam of Bremen's detailed accounts shed light on the intricate exchange between Norse mythology and Christianization, providing invaluable insights into the cultural and religious landscape of medieval Scandinavia.

Adam of Bremen's account is also important to the present study because it provides multiple examples of Christian saints performing acts of weather miracles that take place in Scandinavia. These examples will be further discussed in the chapter on medieval Scandinavian weather miracles and their relation to portrayals of non-Christian weather magic (Chapter 5).

Galdrabækur *and Old English Books of Magic*

Much like other regions in Europe, Scandinavia has its own tradition of books of magic that contain magical spells, runes, and *galdrastafir* (magical staves), called *galdrabækur*. Although there are no surviving examples of *galdrabækur* from the Middle Ages, references to grimoires do exist in medieval Old Norse literature. For example, in *Jóns þáttr biskups Halldórssonar*, there is a scene where Jón Halldórsson reads from one of these books of magic and unintentionally summons a magical storm.[11] While there are no surviving examples of these *galdrabækur* from the medieval period, their inclusion in Old Norse texts demonstrates that, at the very least, a notion of these books of magic was in circulation in Iceland during the Middle Ages. The scene from *Jóns þáttr biskups Halldórssonar* also reveals that these books were connected to weather magic, and are therefore relevant to the present study.

The most famous example of weather-related *galdrastafir* is called *vegvísir* and is included in the Huld Manuscript ÍB 383 4to, which was compiled in the 1800s. Although the symbol does relate to the weather in that it is supposed to guide the user through bad weather, it is not a form of weather magic as defined in this book and therefore, will not be examined extensively in the present study. Instead, this study focuses on the weather-magic staves found in Lbs 2413 8vo, which is a manuscript dated to 1800. In the chapter on weather-magic performances (Chapter 3), I provide an analysis of these staves to determine whether the magic references stem from the medieval period. In this study, I also refer to the *Old English Herbarium*, which is an Old English medical text from c. 1000, which contains weather-magic charms similar to those that are found in the nineteenth-century *galdrabækur*.

Court Records from Early Modern Witch Trials in Finnmark, Norway

Though I maintain a degree of scepticism towards the utilisation of early modern sources in discussions of medieval weather magic, it is pertinent to reference two early modern witch trials that occurred in Finnmark, Norway. These trials bear significance to the present study due to the prevailing argument in previous scholarship that weather magic was among the ritual toolkit of the *noaidi*, or Saami religious ritual performers, despite the absence of direct evidence of such weather-magic rituals before the early modern period. Presumably, these assertions are in part based on the witch trial records from Finnmark. To effectively present these texts and illustrate why they are not reliable sources for reconstructing medieval weather-magic traditions, it is imperative to provide context on the trials and their records.

The witch trials in Norway transpired between 1561 and 1760, constituting the most severe trials in the Nordic countries. The seventeenth century witnessed a multitude of witch trials in Finnmark, which took place between 1601 and 1692, representing, as argued by Rune Hagen, the apex of persecution during peacetime in Norway (Hagen 1999: 43). These trials resulted in the conviction of 137 individuals, 88 of whom received death sentences (Hagen 1999: 44).

The two trials that will be referred to in the forthcoming chapter on Indigenous weather magic (Chapter 4) are those of Anders Poulsen (Poala-Ánde) and Quiwe Baarsen. The first of the two trials to take place is the 1627 witch trial of Quiwe Baarsen, which happened in Hasvåg, in western Finnmark.

The court records indicate that Quiwe Baarsen admitted to frequently conjuring sailing winds for clients. As underscored by Rune Hagen (1999: 58), Baarsen's case is particularly noteworthy as it marks the earliest extant description of the use of a Saami drum in Norwegian legal sources. The trial of Anders Poulsen occurred in Vadsø in 1692. In the court records, Poulsen described extensively the utilisation of a Saami drum, which conferred the ability to forecast the weather and generate favourable winds for sailing. While these trials provide intriguing insights into the beliefs and practices of the early modern period, their reliability as sources for understanding medieval weather magic is tenuous at best. The historical context of witch trials, characterised by fear and the suppression of marginalised communities, complicates their interpretation and renders them problematic for drawing definitive conclusions about pre-modern magical traditions. Thus, while these trials offer potential glimpses into past beliefs and practices, they must be approached with caution and contextualised within broader historical and cultural frameworks.

Organisation of the Book

The first chapter of the monograph examines weather-magic terminology within Old Norse saga literature. Through a deep analysis of weather-magic terms such as *gerningaveðr*, *gerningahríð*, and *galdrahríð*, alongside complementary magic terminology, this chapter reveals the medieval Icelandic understanding of weather magic and its potential ties to non-Christian traditions. Moreover, it demonstrates that weather magic was not merely a literary motif but a social construct embedded within medieval Icelandic society.

Chapter 2 focuses on weather divination and efficatory weather magic. It explores scenes of weather divination and omens in Old Norse saga literature, reflecting a societal interest in controlling destiny and predicting weather patterns crucial for everyday activities like farming and fishing. Through an examination of scenes in Old Norse saga literature that depict performances of efficatory weather magic, this chapter also addresses the problematic categorisation of weather magic into harmful and beneficial categories, emphasising the complexities of Old Norse magic.

The third chapter investigates the various portrayals of weather-magic performance in Old Norse texts, alongside an analysis of potential archaeological evidence. It examines sympathetic magic, speech act theory, and the use of runes and amulets for weather manipulation. Contrary to popular belief, the chapter concludes that weather magic was predominantly performed vocally or through gestures rather than through magical books or leechcraft.

Chapter 4 investigates the complex portrayal of Indigenous weather magic, with a focus on scenes set in Fennoscandia. It examines various textual sources, including Old Norse sagas and historical chronicles, to unravel the complexities surrounding Indigenous weather magic and its practitioners. The chapter highlights the diversity of Indigenous groups in the region, such as the Saami, Bjarmians, and Kvens, challenging the tendency to homogenise these communities. Furthermore, it explores the shared cultural elements of weather magic among Norse and Indigenous communities, revealing a dynamic interplay of beliefs and practices. The chapter also analyses the concept of wind knots and the transmission of magical knowledge between Norse and Saami individuals. Legal and religious texts provide additional insights into the interaction between these communities and the gradual process of Saami conversion to Christianity. Overall, the chapter emphasises the need for a nuanced understanding of weather magic in the different cultures of medieval Scandinavia, shedding light on the interconnectedness and cultural exchange that characterised the region during this period.

The fifth chapter dives into the nuanced portrayal of weather manipulation practices within medieval Scandinavian society, drawing comparisons between non-Christian weather magic and Christian weather miracles found in Old Norse texts. It challenges the notion of a clear dichotomy between these practices, suggesting instead a more nuanced interaction. Examples from sagas like *Ólafs saga Tryggvasonar* and *Jóns saga ins helga* illustrate how Christian authorities overturned non-Christian weather magic, indicating a belief in the superiority of Christian divine power. The text explores cultural loan shifts, proposing that local church authorities may have consciously replaced reliance on non-Christian magic with Christian 'counter-magi' figures capable of weather manipulation through God's power. Notably, the chapter highlights continuities in the performance of weather magic, such as the emphasis on vocal performance, while also emphasising uniquely Scandinavian motivations for weather miracles, such as agricultural success and maritime safety. Overall, the chapter underscores the complexity of these practices and their adaptation within a changing religious and cultural landscape in medieval Scandinavia.

The final chapter explores the role of Óðinn as a weather-magic practitioner, challenging previous scholarly emphasis on Þórr as the primary Norse weather deity. Through an extensive analysis of literary and poetic representations, this chapter highlights Óðinn's association with weather magic in the Scandinavian Middle Ages.

Notes

1 For example, *Myths and Magic in the Medieval Far North: Realities and Representations of a Region on the Edge of Europe* (Figenschow, Holt, and Tveit 2021), *Myth, Magic and Memory in Early Scandinavian Narrative Culture: Studies in Honour of Stephen A. Mitchell* (Glauser and Hermann 2021), *The Norse Sorceress: Mind and Materiality in the Viking World* (Gardeła, Bønding, and Pentz 2023), *Performing Magic in the Pre-Modern North* (Hemphill, Richardson-Read, and Wang 2024).

2 The term 'Indigenous' is used throughout this monograph as it is defined in the UN's ILO Convention 169. Here, the term is defined as 'peoples in independent countries who are regarded as Indigenous on account of their descent from the populations which inhabited the country, or a geographical region to which the country belongs, at the time of conquest or colonisation or the establishment of present state boundaries and who, irrespective of their legal status, retain some or all of their own social, cultural and political institutions' (ILO 169).
3 Denoted by Terry Gunnell (2014), the 'goatskin twirl' is a form of sympathetic weather magic in Old Norse literature where a magic practitioner twirls or waves a goatskin to bring about weather changes, such as creating fog or storms. 'Magical mooning' is another type of sympathetic magic, also named by Gunnell (2014), involving a practitioner exposing their bare buttocks to manipulate the natural world, often with the intent of overturning the landscape or driving enemies mad.
4 A parallel can be seen in Finnish tradition, where the nineteenth-century *Kalevala* likewise preserves elements of mythological magic through poetry. Louhi, ruler of Pohjola, is portrayed as a storm-working figure who summons tempests against her enemies. This recalls Old Norse poetic depictions of storm-associated beings such as Hræsvelgr, the giant eagle whose wingbeats generate the winds, and Þjazi, a *jötunn* in eagle form linked to violent natural forces. Louhi herself takes the shape of a great bird in the *Kalevala*, which further underscores the thematic resonance between avian figures and storm-magic in northern poetic traditions. The many Old Norse saga accounts of women raising winds provide additional parallels. Yet the *Kalevala* is a nineteenth-century compilation from oral sources and cannot be treated as medieval evidence. Its value lies instead in showing how the association of magic, poetry, and myth with magical storm-making persisted across northern traditions, even if no direct historical connection links it to medieval Scandinavian material.
5 See Mitchell 1991 for further discussion on the *fornaldarsǫgur*.
6 Andrew Dennis, Peter Foote, and Richard Perkins use the terms 'witchcraft' as a translation for *gørningar* and 'black sorcery' for the term *fordæðuskapr*. However, these are not direct translations of these terms and the use of the words appears to be made as a choice to distinguish the levels of severity between the two types of magic and their punishments. Magical terminology will be discussed further in the first chapter of the present study.
7 The old man portrayed in this scene is most likely Óðinn, who often comes to heroes during sea battles in the form of an old man, who is able to dispel magical storms with his own weather magic. This motif will be further discussed in Chapter 6 about Óðinn's skills in weather magic.
8 Erik's ability to control the weather is explained in the *Swedish Prose Chronicle* from the 1450s, where he is able to call a fair wind in whichever direction he turns his hat (Saxo ix.4.33: 657).
9 It is interesting to note that in the Scottish Hebrides, there is a similar folklore tradition, where the mermen referred to as the 'blue men of the Minch' or 'storm kelpies' also bring storms at sea.
10 Perkins notes that although there are discrepancies in how the text has been interpreted and transcribed by previous scholars, Dudo's account can 'safely be interpreted as demonstrating that human sacrifice was offered to Thor in expectation of assuring favourable winds at sea' (Perkins 2001: 20).
11 This scene is further contextualised and analysed in the chapter on weather-magic performance (Chapter 3).

PART I

Conceptualising Weather Magic

1

LEXICON OF STORMS

Weather-Magic Terminology

From a lexicological standpoint, examining medieval Scandinavian magic can be a difficult task due to the variety of terminology that thirteenth- and fourteenth-century sources used to convey the ability to magically manipulate one's surroundings (Raudvere 2002; Meylan 2014). It is impossible to provide a precise classification of Old Norse magical terminology because the uses and meanings of these words within written sources are inconsistent. A single word does not always convey the same implication each time it is used, nor are the performers of magic consistently denoted using the same descriptive terminology. Trying to force a modern sense of taxonomies onto medieval Scandinavian magic is unrealistic when the Old Norse language surrounding these concepts lacks concrete and unchanging definitions (Ármann Jakobsson 2017: 62). That being said, Catharina Raudvere (2002: 80) proposes that the contradictions surrounding magical terminology within the Old Norse corpus are not a result of confusion on behalf of the Norse but are perhaps due to magic being used to explain a variety of different aspects within medieval Norse life. Other scholars (Hastrup 1990; Mitchell 2011) suggest that Old Norse magic touched all aspects of life in one way or another.

Given this complexity, it is clear that medieval Scandinavian magic cannot be easily categorised or defined by a single term. This lack of consistent terminology in Old Norse sources reflects the diverse nature of magical practices and their varied roles within Norse society. Consequently, any attempt to impose a uniform classification on these practices risks misrepresenting the complexity of magical beliefs and their social roles. Recognising the fluid and context-dependent nature of these terms allows for a more nuanced understanding of how magic was perceived and practised in medieval Scandinavia.

DOI: 10.4324/9781003746058-3

This nuanced approach is essential as we consider the broader implications of using modern, generic umbrella terms to describe these practices.

Using modern, generic umbrella terms like 'witchcraft' and 'witch' to describe Old Norse magical practice and its practitioners is problematic. These words carry pre-established connotations from a wide range of contexts that are unrelated to medieval Scandinavia (Burke 1992: 87; Mitchell 2011). To avoid these problematic undertones, Raudvere uses the emic word *trolldómr* to 'indicate the notions, rituals, and social interactions in the Old Norse traditions relating to conceptions about the influence certain persons had – by innate qualities or through skill – on the world surrounding them' (Raudvere 2002: 87). Raudvere further explains that the concept of *trolldómr* focuses on the individual's ability to influence others and the surrounding environment, with either positive or negative intentions. However, for this study I have chosen to use the term 'magic' rather than *trolldómr*, or another emic umbrella term. This choice has been made because *trolldómr* is often found in contexts where magic and its practitioners are portrayed in a negative light (Meylan 2014: 25). The term is derived from the word *troll* (Cleasby and Guðbrandur Vigfússon 1874: s.v. troll), which can often be applied to beings or ideas that evoke a sense of 'alienness' or 'other' (Ármann Jakobsson 2017: 19). This connection thus portrays *trolldómr* and its practitioners as hazardous, resulting in the term being used in a pejorative manner. Furthermore, the use of *trolldómr* as an umbrella term is problematic in that the Old Norse literature analysed throughout this study does not use one specific term to indicate magical practice. Instead, a variety of terminology that designates magical practice is found throughout the Old Norse corpus.

In a similar manner, I have chosen to use the terms 'practitioner' or 'performer' rather than employing an Old Norse term to indicate those who practise magic. Again, this decision has been made as a result of the diverse terminology which is used to describe magically-skilled individuals throughout Old Norse literature. Male practitioners of magic are mostly referred to as *galdramaðr*, *vitki*, *skratti*, or *trollmaðr*, while female practitioners are typically called *gýgr*, *seiðkona*, *spákona*, *trollkona*, and *vǫlva* (Raudvere 2002: 89). Much previous scholarship (Morris 1991; Jochens 1996; Jóhanna Katrín Friðriksdóttir 2013) has focused on the female role in magical performance. While it is true that some magical practices in an Old Norse context, like *seiðr* and divination, have been interpreted by scholars as female roles in the past, several scholars (Gísli Pálsson 1991; Dillmann 2006; Merkelbach 2019) have noted that an almost equal number of male and female practitioners can be found in the *Íslendingasǫgur*. In Old Norse literature, weather magic is represented as a universal practice, regardless of the performer's gender.

In her article investigating the narrative role of magic in the *fornaldarsǫgur*, Aðalheiður Guðmundsdóttir (2015) argues that episodes of magic, in these legendary sagas specifically, are most commonly used to fulfil certain narrative functions. She highlights that magic is often employed throughout this specific genre to provide the hero with an opponent, to emphasise the hero's moral standing, present non-Christian traditions, or provide enhanced entertainment (Aðalheiður Guðmundsdóttir 2015: 52). Aðalheiður Guðmundsdóttir's article is one of many recent academic studies that examines Old Norse terminology to create a clearer picture of the medieval Icelandic understanding of magic (e.g. Raudvere 2002; Dillmann 2006; Heide 2006; Meylan 2014). A similar understanding of the weather-magic motif and its traces of an older, pre-Christian tradition can be developed by investigating the terminology that scribes used when depicting occurrences of weather magic in the Old Norse source material. By examining the weather-magic terminology that is used throughout the entire Old Norse corpus, this section aims to present a comprehensive understanding of weather magic as it was understood in the Scandinavian Middle Ages.

Before examining the Old Norse terminology that specifically relates to weather magic, it is necessary to address the problematic way that the terms 'supernatural' and 'paranormal' have been used throughout the field of Scandinavian Studies, as well as the issues of classifying magic into such a category. Until recent scholarship, 'supernatural' was often employed throughout research when referring to unknown phenomena that occur in the Middle Ages. However, new scholarship has highlighted the problem that using the term 'supernatural' poses (Ármann Jakobsson 2017; Merkelbach 2019). Applying the word 'supernatural' to magical abilities implies that the concept was somehow above nature or not of this world. In the Middle Ages, having magical skills was not necessarily viewed as existing above nature; rather, they were seen as extraordinary but still part of the natural world. To counteract this issue, Ármann Jakobsson (2017: 22–23) suggests replacing 'supernatural' with the modern term 'paranormal'. He explains that because the focus should be fixed onto the human experience and human society, the stem '-normal' in paranormal is preferable to the stem of supernatural, '-natural'. This preference is a result of nature being envisioned by modern audiences as 'all that is distinct from humanity' (Ármann Jakobsson 2017: 22).

Furthermore, Ármann Jakobsson (2017: 22) emphasises that modern audiences rarely take medieval concepts that are classified as 'supernatural' seriously. These ideas are simply seen as not relatable, and therefore unbelievable, to the modern reader. As a result, he suggests that replacing 'supernatural' with the modern word 'paranormal' is ideal as 'paranormal' is rarely used in a medieval sense. To present-day audiences, 'paranormal' is associated with concepts like aliens, ghosts, and poltergeists. Although many of us are sceptical as to whether these otherworldly beings actually exist, they are concepts that are still taken seriously by much of the modern world. By applying this modern concept to the Middle Ages, Ármann Jakobsson suggests that the use of 'paranormal' does not automatically suggest something medieval, and 'thus it cannot be taken for granted, dismissed as a traditional or conventional term that can be safely deployed without scrutiny' (Ármann Jakobsson 2017: 22).

However, the term 'paranormal' also poses a problem for the study of Old Norse magic because, like 'supernatural', it implies a separation from what medieval Scandinavians would have considered a part of everyday life. In reality, magic as a concept was fully integrated into Old Norse society and was taken very seriously, even after the region's conversion to Christianity. Additionally, I disagree with the notion that modern audiences take the term 'paranormal' more seriously. Given the rise of modern entertainment revolving around the paranormal, such as movies, TV shows and books, modern audiences have developed preconceived notions about the topic, often viewing it as fantastical rather than factual.

Therefore, I have chosen to avoid the use of terms like 'supernatural' and 'paranormal' in this study. Instead, I use the words 'magic' and 'magical'. Although the term 'magic' itself can be problematic due to its broad and varied connotations, it is the most appropriate term for this book as it directly addresses the cultural and historical practices being examined. 'Magic' accurately reflects the medieval Scandinavian worldview where such practices were part of the natural order and everyday life. By using 'magic', I aim to remain true to the context of weather magic in medieval Scandinavia, emphasising its significance and integration into the societal and cultural milieu.

In *Defining Magic: A Reader*, Bernd-Christian Otto and Michael Stausberg (2013) examine the complex and highly debated term 'magic', as it is used within the academic field of the history of religion. Similar to the issues noted above regarding 'paranormal' and 'supernatural', Otto and Stausberg (2013) emphasise that the term 'magic' is also problematic due to historical biases and that the broad term has not achieved a universally accepted academic definition or theoretical framework. They point out that throughout history, the word 'magic' has been employed as a pejorative term to critique or distinguish certain practices from the realms of religion, science, and rationality. The term has been used to bring doubt to practices deemed irrational or illegitimate, often positioning them as inferior counterparts to structured religion or empirical science. The tendency to place magic into a pejorative category reflects the broader socio-cultural efforts to marginalise and devalue practices that do not conform to mainstream religious or scientific paradigms. Despite acknowledging the broad and often negatively biased connotations associated with the term 'magic', I argue that it is a less problematic descriptor than 'supernatural' or 'paranormal'. Unlike these latter terms, 'magic' is more universally recognised and understood by modern audiences, facilitating a clearer communication of the concepts being examined. The term 'magic' directly engages with the cultural and historical practices central to this study, whereas 'supernatural' and 'paranormal' risk alienating these practices by implying a separation from the natural world or suggesting an element of modern fantastical entertainment.

Valerie I.J. Flint offers a concise yet comprehensive definition of magic as 'the exercise of preternatural control over nature by human beings, with the assistance of forces more powerful than they' (Flint 1991: 3). For the purposes of this study, Flint's definition is particularly useful, though it requires a slight adaptation. In the Old Norse context, it is not only humans who wield magical skills; the gods themselves are often depicted as proficient in magic, possessing knowledge and abilities that parallel those of human practitioners. This inclusion broadens the scope of our understanding, integrating divine figures into the framework of magical practice.

In *Witchcraft and Magic in the Nordic Middle Ages* (2011), Stephen Mitchell provides another layer of understanding by defining magic within medieval Scandinavian societies as practices believed to manipulate the world through what he describes as supernatural means. Mitchell emphasises that magic in these contexts was not merely about the manipulation of the natural world; it was a concept that was deeply incorporated with societal norms, gender roles, and legal structures. For example, episodes of weather magic in Old Norse texts reveal that such practices were felt to be embedded in the fabric of everyday life and cultural narratives. These instances represent magic, not as an isolated activity or separated from the natural word, but as a practice deeply intertwined with the social and cultural milieu of the time.

The many episodes of weather magic found in Old Norse literature illustrate that magic was seen as a pervasive and integral part of medieval Scandinavian life. These narratives often depict magic as a tool wielded by men, women and gods alike. By examining these episodes, scholars can discern that weather magic, and magic in general, was not an anomaly but a recognised and significant aspect of life, embedded in the cultural consciousness and everyday practices of medieval Scandinavians. This study aims to highlight the deep integration of magic into the societal and cultural narratives of the time, demonstrating its importance and ubiquity in medieval Scandinavian society.

Knowledge

The ability to perform magic in medieval Scandinavia was linked to the practitioner possessing a specific type of heightened knowledge, which allowed them to exert control over their surroundings. The importance of having a particular understanding of magic is often highlighted through the use of such Old Norse words as *fjǫlkunnigr* or *margkunnigr* to describe the performer, and *fjǫlkynni* to describe the practice itself (Merkelbach 2019: 127; Meylan 2014: 32; Dillmann 2006: 194–212; Raudvere 2002: 88).

The second part of these compounds are derived from the Old Norse word *kunna* (Merkelbach 2019: 127; Meylan 2014: 30), meaning 'to know' or 'to understand' (Cleasby and Guðbrandur Vigfússon 1874: s.v. kunna). As Mitchell (2019: 138–139) points out, this nuanced vocabulary used to describe magical knowledge reflects a cultural framework in which magical expertise was tied to elevated, often elite forms of understanding. The relation between magic and an individual's knowledge is often highlighted in the narration of weather-magic performances by describing the weather-magic performer as *fjǫlkunnigr*. One such example is demonstrated in *Víglundar saga*, when the character Kjǫlvur, a woman living at Hraunskarð, summons a storm by waving her veil towards the eastern sky (ÍF 14: 82; CSI 2: 421–422). Prior to Kjǫlvur casting her magic, the narrative informs the audience that 'hun var fjǫlkunnig mjǫk' (ÍF 14: 82) [she was very skilled in magic]. Here, the use of *fjǫlkunnigr* highlights the connection between Kjǫlvur's ability to manipulate the weather and her specific understanding of magic.

While the term *fjǫlkunnigr* is often associated with individuals exercising their knowledge in magic with malicious intentions, it is important to recognise that this is not universally the case. In Kjǫlvur's scenario, the narrative does present her as a malevolent character. Following the assertion that she is *fjǫlkunnig mjǫk*, the saga continues to describe her as 'ok at öllu illa fallin, hardla óvinsæl við alþýðu manna' (ÍF 14: 82) ['completely given to evil, so that she was quite unpopular with people everywhere' (CSI 2: 421)]. Her intentions of mustering the storm to create a deadly situation for Víglundr, the saga's main protagonist, indeed align with this negative portrayal. Prior to Kǫlvur's weather magic taking place, Þorbjǫrg and her two sons, Jǫkull and Einarr, pay Kjǫlvur a sum of a hundred pieces to kill Víglundr and Trausti by any magic that she chooses (ÍF 14: 82; CSI 2: 421). The transaction between Þorbjǫrg and Kjǫlvur reveals Kjǫlvur's willingness to use her magical knowledge to cause harm and destruction.

However, the connection between having particular knowledge to perform weather magic and the connotations of such skills can be more complex, and the use of the term *fjǫlkunnig* to designate the performer does not always indicate that they are using their knowledge with malintent. In Chapter 35 of *Laxdœla saga*, Kotkell and his family are described as 'mjǫk fjǫlkunnig ok enir mestu seiðmenn' (Kålund 1889–1891: 34–35) ['skilled in [magic] and accomplished magicians' (CSI 5: 50)]. While they call a deadly storm near Skálmarnes with the intention to destroy the ships belonging to their enemy Þórðr Ingunnarson, not all uses of weather magic are inherently malevolent or destructive. The notion of beneficial versus harmful weather magic will be explored in depth in the second chapter of the present study.

The use of weather magic extends beyond Icelandic settings, and is not always used to describe Norse practitioners. Indigenous characters are also portrayed as having mastered this special knowledge of manipulating the weather. In *Ólafs saga helga*, King Ólafr sails north to raid the coastline and encounters an Indigenous community. After the Indigenous community attack the king and his men, they invoke this magical knowledge to further defend themselves: *Þeir Finnar gerðu um nóttina œðiveðr með fjölkyngi ok storm sjávar* (ÍF 27: 11) ['During the night the [Saami] caused furious weather by magic, and a storm at sea' (*Heimskringla* II: 8)]. This connection between the Saami and their understanding of magic is highlighted through the use of terminology when King Ólafr's good fortune outmanoeuvres the *fjölkyngi Finna* (ÍF 27: 11), or 'magic of the [Saami]' (*Heimskringla* II: 8).

Overall, while having a specialised knowledge of weather magic can be portrayed as a tool for malevolent intent, it is essential to understand its broader and more nuanced applications within Old Norse texts. The weather-magic scene above from *Ólafs saga helga* will be examined further in the following chapter, as an example of how it is complicated to divide magic into the harmful versus non-harmful binary. In these texts, scenes of practitioners employing their specialised knowledge of weather magic, using the designations *fjǫlkunnigr* or *margkunnigr*, could be used by the narrator to serve various purposes, not all of which were harmful, reflecting the complex nature of magic in medieval Scandinavian society.

Magical Storms

Within the Old Norse corpus, there are three terms that are specifically associated with storms created by weather magic: *gerningaveðr* and *gerningahríð*, meaning 'storm/weather raised by magic', and *galdrahríð*, meaning 'magic storm'.

TABLE 1.1 Weather-Magic Terminology, Old Norse. The table has been made by the author, adapted and added to from data retrieved from the *Dictionary of Old Norse Prose*

Term	*Forms (occurrence)*	*Saga Group (occurrence)*	*Text*
gerningaveðr – noun, neuter	*gerningaveðr* (2); *gerninga veðr* (1); *gíörnínga uedur* (1); *giorninga veðr* (1)	*Íslendingasǫgur* (4); *konungasǫgur* (1)	*Hallfreðar saga vandræðaskálds; Norna-Gests þáttr; Harðar saga ok Hólmverja; Vatnsdœla saga*
gerningahríð – noun, feminine	*gerníngahríð* (1)	*fornaldarsǫgur* (1)	*Gǫngu-Hrólfs saga*
galdrahríð – noun, feminine	*galldra hryá* (1)	*fornaldarsǫgur* (1)	*Hrólfs saga kraka*

In Old Norse literary sources, magic is understood as an act performed by a practitioner, which is demonstrated by the use of terms such as *gerningar* (Dillmann 2006: 214–216; Merkelbach 2019: 125), which is derived from the verb *gera*, meaning 'to do' (Cleasby and Guðbrandur Vigfússon 1874: s.v. gera); the word *gerning* means 'act' or 'deed' (Meylan 2014: 32). When the word takes the plural form *gerningar*, the term becomes one that designates magical practices or magic in general (Dillmann 2006: 214). From *gerningar*, weather-magic-related compounds like *gerningaveðr* and *gerningahríð*, both meaning a storm conjured with magic, are used to describe particular deeds rather than abstract discussions of magic (Meylan 2014: 32).

Gerningaveðr

The texts that utilise these *gerningar*-compound terms normally provide very little description of the magical performance itself. Instead, the narrators rely on the audience to understand the term and its magical implications. As noted above, Meylan suggests that these terms are used to emphasise the actual deed of magic. However, the typical lack of description following these terms fails to reveal to modern audiences what the medieval Icelanders' understanding of the practicalities surrounding the action was. An investigation into the cultural understanding of weather-magic performances in medieval Scandinavia will be presented in the third chapter of this study. Out of the three weather-magic-specific terms listed above, *gerningaveðr* is the compound word used to portray weather magic most frequently in Old Norse literature. That being said, there are only five known examples of *gerningaveðr* that exist in the surviving medieval Icelandic sources. *Gerningar* indicates a magical act while *veðr*, the Old Norse term for 'weather', links the preceding magical action to weather magic. In Old Norse, the word *veðr* is often used to indicate bad or extreme weather (Cleasby and Guðbrandur Vigfússon 1874: s.v. veðr).

When examining the use of *gerningaveðr* to indicate weather-magic performances in Old Norse literature, the majority of occurrences are found within the *Íslendingasǫgur*, with the exception of one *þáttr*. The first known use of this word occurs within the Möðruvallabók version of *Hallfreðar saga vandræðaskálds* (AM 132 fol.),[1] a manuscript that is normally dated between c. 1330 and 1370 (ÍF 8). After wintering at his brother's abode, Hallfreðr heads north and comes across a magical blizzard: *Ok um verit er hann fór norðr þa rak a firir þeim hrið. Hallfreðr kvað þat vera gørningaveðr* (ÍF 8: 187) [And in the spring he was travelling north a magical blizzard started up ahead of them]. In the text, the narrator using the term *gørningaveðr* indicates that it is a storm raised by magical means. The storm directs Hallfreðr and his travel companions to an enclosure, where they clash with Már of Másstaðr and his twenty men. In this first example, along with the majority of examples that will be described below, the text does not offer further explanation of the magical action surrounding the use of the word *gerningaveðr*. The manuscript omits any reference to the performer of the weather magic, nor does it provide a description of the performance. The choice to rely solely on *gerningar* to indicate magical intervention indicates an assumption that the reader or listener is already familiar with the word and its magical implications. On the other hand, the term alone does not give any indication of the medieval Icelandic understanding of how this type of magic could have been performed.

The second known use of *gerningaveðr* is found in *Norna-Gests þáttr*, a tale about memories of ancient times that is preserved in the manuscript AM 62 fol., which is normally dated between c. 1375 and 1400 AD (Ólafur Halldórsson 2000). However, the story itself is preserved in multiple manuscripts and scholars usually date the work to around 1300 (Harris and Hill 1989: 105). In the narrative, Sigurðr and his men encounter a storm at sea when sailing to meet the sons of Hunding in battle. The narrator's word choice in this passage indicates that the tempest is summoned by magic and the performance is then attributed to the sons of Hunding: 'Þa fingv ver gerninga veðr stor. Þat kendv menn Hvundings s(onvm)' (Ólafur Halldórsson 2000: 24) [Then we got a large storm raised by magic, and people knew that it was the sons of Hunding (own translation)]. Fleeing from the storm, Sigurðr and his crew sail along the shore and meet a mysterious old man who is understood to be Óðinn, who comes on board their ship and calms the storm. This scene is also portrayed in *Reginsmál* (Neckel and Kuhn 1962: 177; Larrington 2014: 151) and *Vǫlsunga saga* (Guðni Jónsson and Bjarni Vilhjálmsson 1943–1944), although in both of these sources, the creation of the storm is not attributed to weather magic. In *Reginsmál*, the storm is portrayed as a *storm mikinn* (Neckel and Kuhn 1962: 177), or a 'great storm', with no magical inference. In *Vǫlsunga saga*, the abrupt turn in weather is narrated: *þá kom á veðr mikit með stormi* (FSN 3: 156) [then came great weather with a storm]. This depiction also gives no indication that the storm was called by magic. Notwithstanding, in all of the portrayals mentioned above, Óðinn is responsible for calming the storm.[2]

The scene in *Norna-Gests þáttr* is the earliest example of the word being used where the performers of the magic act are also revealed. Yet it is attributed to a group of people rather than a specific individual. The narrative omits any further description of the sons of Hunding's knowledge of magic, nor does the content include an explanation of the performance. Again, the lack of additional information and description surrounding the use of *gerningaveðr* suggests that using the magical terminology alone was enough to inform the audience of the action and its magical undertones, but similar to the previously described examples, it does not provide any insight into the medieval understanding of how weather magic was performed.

Another weather-magic performance that includes the use of *gerningaveðr* is found in *Harðar saga ok Hólmverja*. This *Íslendingasaga* is preserved within the manuscript AM 556 a 4°, which is dated between c. 1475 and 1500 (Hast 1960). In this scene, the weather magic occurs over winter when Hörðr and fifteen of his men steal the sheep that belong to Indriði. The unnatural weather is described in the phrase: '*Þá giörde snæfall míkit. ok gíörnínga uedur mote þeim*' (Hast 1960: 163) [There was heavy snowfall and magical weather against them]. The text includes no indication of who performs the magical act and for what purpose the storm is conjured, other than possibly inconveniencing Harðr's efforts. Once again, the text relies on the term alone to inform the audience of a magical manipulation of the weather without providing further details of how the act is actually performed.

The final two known examples of *gerningaveðr* are found in *Vatnsdœla saga*, an *Íslendingasaga* that was probably written in the late thirteenth or early fourteenth century. *Vatnsdœla saga* was most likely present in the now-lost Vatnshyrna manuscript, which was destroyed in the Copenhagen fire of 1728. Some of the sagas that are thought to be from Vatnshyrna are included in what is known as the Pseudo-Vatnshyrna,[3] which is dated around 1400. However, there is no way of ensuring whether the Vatnshyrna version of *Vatnsdœla saga* contained the same wording that is used in the Pseuedo-Vatnshyrna fragments. The first time that *gerningaveðr* is used in *Vatnsdœla saga* is found in the later recension AM 396 fol., which is dated c. 1675–1700. *Gerningaveðr* is used when Faxi-Brandr and Jǫkull claim that Helga at Borg is at fault for an unnatural snowstorm by saying, '*Þeir... kváðu þetta gørningaveðr verit hafa ok kenndu þat Helgu frá Borg*' (ÍF 8: 92) [Faxi-Brand and Jokul claimed that there had been a storm produced by weather magic and blamed it on Helga at Borg (own translation)]. The text also reveals that Helga's purpose for raising the storm is to prevent a duel that was supposed to take place between her husband, Bergr, and Jǫkull.

This scene is interesting because it is an example of a weather-magic compound where the individual performer is stated, alongside the additional information regarding the performer's magical knowledge. Prior to casting the blizzard, Helga is described as an imposing woman who is gifted with divination skills and is *margkunnig um flesta hluti* (ÍF 8: 89), or 'wise in magical ways' (own translation). When Helga summons a magical storm, the audience is already aware that she is capable of performing magic. That being said, the text does not expand on the practicalities of how Helga performs the weather magic. Outside the use of *gerningaveðr*, there is no additional indication of magic nor a description of the skills used to create the blizzard.

Although this source cannot provide insight into how an act of weather magic might have been performed, it can reveal a possible narrative function of this type of magic. One can argue that the above example of weather magic was implemented to fulfil the narrative role of providing a female character with a certain autonomy. In this specific situation, the act of weather magic is used as a method for Helga to exert control over the outcome of a situation, which would normally be determined by men (Jóhanna Katrín Friðriksdóttir 2013: 54). This outcome is presented as negative in the text because Bergr is humiliated for not showing up to the duel.[4] A similar depiction of this episode can be found within *Finnboga saga ramma*, although this second portrayal lacks the inclusion of specific weather-magic terminology. This episode tells the same story as the *Vatsndæla saga* scene described above. However, unlike *Vatnsdœla saga*, *Finnboga saga ramma* portrays Bergr as the aggressor rather than the hero. In this version, it is Bergr's partner, Dalla, who utilises weather magic to prevent Bergr from participating in a duel. In both *Vatnsdœla saga* and *Finnboga saga ramma*, Dalla and Helga fail to convince Bergr to avoid the altercation with words alone. Instead, they are forced to use their magical knowledge to prevent the violent encounter. The perceived morality of Helga and Dalla's use of weather magic will be explored in the following chapter on efficatory magic.

The final known use of the *gerningaveðr* compound also occurs in the later recension of *Vatnsdœla saga*, in a similar situation where it is implied that the magic is performed to prevent a duel. However, the performer of the weather magic and their motivation to stop the duel is not stated in this second instance. In the episode, Þórarinn challenges Starri to a duel. In a turn of events, they are unable to make the journey due to the harsh weather resulting from magic:

> þá er Þórarinn illi skoraði á hann til hólmgǫngu, at Úlfheðinn fór með honum til hólmstefnunnar, ok í þeiri ferð gerði at þeim veðr illt, ok ætluðu þeir vera gerningaveðr.
>
> *(ÍF 8: 127)*

> When Thorarin the Evil challenged Starri to a duel, Ulfhedin accompanied him to the duelling place and on the journey the weather turned foul, and they believed it was a storm conjured by magic.
>
> *(own translation)*

The use of weather magic in this scene does not have any considerable implication on the plot, nor can it be interpreted as an act stemming from good or bad intentions. Because their travelling companion, Barðr, has the knowledge to perform a spell that will counteract the magical weather (this act will be discussed in more detail on p. 41), Þórarinn and Úlfheðinn are not delayed for long. Once again, outside of the use of *gerningaveðr*, no further indicators are provided in the text to inform the audience that the weather was due to magical manipulation. Although the performer remains unknown, it seems that weather magic is used once again in an attempt to avoid what would have been a violent encounter.

Gerningahríð

The second weather-magic compound, *gerningahríð*, combines *gerningar* with the Old Norse term *hríð*. Similar to *gerningaveðr*, *gerningar* is the root that is used to link the expression to a magical act. The addition of *hríð*, meaning storm or snowstorm in Old Norse, results in a compound, which indicates a magical manipulation of the weather. Unlike the use of *-veðr*, the stem *-hríð* concretely implies the occurrence of bad or extreme weather. The only known use of *gerningahríð* is found in *Gǫngu-Hrólfs saga*, which is normally dated between c. 1500 to 1525 AD.

In this legendary saga, two of the key characters, Hrólfr and Vilhjálmr, are challenged by King Eirekr of Russia to enter King Hreggviðr's burial mound and retrieve a dead king's armour. On their journey to the enchanted tomb, Hrólfr and Vilhjálmr encounter a blizzard so strong that Hrólfr suspects it must have been summoned by magic: 'Þat þóttist Hrólfr skilja, at hríðin mundi hafa drepit sendimenn konungs, ok þetta mundi gerninga hríð vera' (FSN 3: 279) [Hrolf realised that the blizzard must have killed the king's men, and that it must be a storm sent by magic (Waggoner 2014: 193)]. This blizzard that Hrólfr attributes to magic is described as so severe that large oak branches fall from the trees and the lightning and thunder is so strong that Hrólfr's magical cloak is the only thing that is preventing his death.

Once again, there is no description of the actual magical performance. Yet in this instance, the use of weather-magic terminology varies from the previous examples that include *gerningaveðr* because rather than solely relying on terminology to indicate the act of magic, *Gǫngu-Hrólfs saga* sets the scene by including a detailed account of outcome caused by the use of weather magic. The audience is also informed that the magical act is caused by two of the Russian king's champions, Sörkvir and Grímr Ægir (Waggoner 2014: 194). When Hrólfr and Vilhjálmr first arrive in King Eirekr's realm, Grímr Ægir is described as possessing skills in spells and magic (Waggoner 2014: 190). Having this knowledge means that the reader or listener is already aware of the potential inclusion of magical events. That being said, the narrative's focus is on the results of the weather magic, rather than providing any description of the actual performance.

Nevertheless, the additional details surrounding the event and the awareness of the performer does not mean that whoever wrote down the saga expected that the audience would be unfamiliar with weather magic. *Gǫngu-Hrólfs saga* is an intriguing source because its extant texts contain various *apologiae*[5] inserted within the text to defuse the audiences' scepticism of the many magical events throughout the narrative. The recurring inclusion of *apologiae* within the Old Norse saga tradition suggests that the scribes were criticised by their contemporary audiences for the use of magical elements, implausible content, or variations within the narrative (Kalinke 1985: 318).

The narrator of *Gǫngu-Hrólfs saga* argues that when committing these stories to manuscript, one has the obligation to truthfully convey whatever they have seen or heard, regardless of their personal beliefs.

In the first *apologia* that is situated at the beginning of the saga,[6] the scribe specifically addresses their responsibility to include occurrences of magic, and uses an example of individuals who can manipulate the elements. The writer continues their argument on the importance of including these potentially questionable subjects by saying:

> svá ok eigi síðr um konstir eðr huklaraskap ok mikla fjölkýnngi. Þá þeir seiddu at sumum mönnum æfinliga ógæfu eðr aldrtila, en sumum veraldar virðing, fjár ok metnaðar. Þeir æstu stundum höfuðskepnur, en stundum kyrrðu, svá sem var Oðinn eðr aðrir þeir, er af honum námu galdrlistir eðr lækningar.
>
> *(FSN 3: 237)*

> This is no less the case concerning trickery, or wizardry and mighty magic, when they conjured up eternal misfortune or loss of life for some, and for others, worldly reputation, riches, and honors. Sometimes they stirred up the elements and sometimes calmed them down – men such as Odin was, and the others who learned spellcraft or healing from him.
>
> *(Waggoner 2014: 170)*

Although this passage could be interpreted as a way to explain weather magic to an audience that is unfamiliar with the concept, or an audience that is familiar with the concept but is sceptical of this type of magic, the number of times the weather-magic motif occurs within the Old Norse corpus makes it unlikely that this is the first time the scribe or the audience is encountering this magical motif.

It is contestable whether medieval audiences actually believed in the use of weather magic. Stephen Mitchell proposes that the existence of *gerningaveðr*, *gerningahríð*, and *galdrahríð* within Old Norse vocabulary alone hints that the concept of weather magic may have been one of significance in medieval Scandinavia (Mitchell 2013: 65). Yet compared to the magnitude of magical terminology used in Aðalheiður Guðmundsdóttir's study, there are only a few known terms specific to weather magic and all of the known uses of these words are limited to manuscripts that range from the fourteenth to the seventeenth century. The seven known occurrences of these terms within the entire Old Norse corpus cannot alone provide enough evidence to prove that weather magic was significant in medieval Scandinavia. Nevertheless, the sheer existence of these terms combined with the many occurrences of weather-magic motifs within the corpus means we can deduce that there was an established tradition of weather magic ingrained in popular Scandinavian belief.

Prior to the known uses of weather-magic terminology outlined above, similar acts of weather magic are portrayed without the inclusion of *gerningar* compounds. An example of such is found in the previously discussed scene from *Laxdœla saga* (Chapter 35), an *Íslendingasaga* that is normally dated to the thirteenth century. The earliest surviving manuscript of this saga is a fragment from c. 1250. However, the earliest complete version of this saga is found in AM 132 fol., the same manuscript that contains the previously mentioned scene in *Hallfreðar saga vandræðaskálds*, which is normally dated between c. 1330 and 1370. The weather-magic episode is found in the full manuscript, rather than the earlier fragment. In this scene, Kotkell's family of magic practitioners conjure a storm to sink the ships of Þórðr Ingunnarson. At the Alþingi, Þórðr charges Kotkell and his family with magic and theft, a crime punishable by full outlawry. After Kotkell breaks the news to his family, they take revenge with their knowledge of magic. The text narrates: 'Síðan lét Kotkell gera seiðhjall mikinn; þau fœrðusk þar á upp ǫll; þau kváðu par harðsnúin frœði, þat váru galdrar. Því næst laust á hrið mikilli' (ÍF 5: 99) [Kotkell then prepared a high platform for [magic] which they all mounted. They chanted powerful incantations, which were [magic]. A great blizzard came up (CSI 5: 50)]. This episode is one of the many examples of weather magic that is depicted without the use of weather-magic specific terminology and it highlights a similar idea of conjuring magical storms. Yet unlike most texts including *gerningar*-compound terms, this text provides a description of the actions surrounding the weather-magic performance, which will be discussed further in Chapter 3 that analyses the actions and performance of weather magic. As explained above, the later sources that do include weather-magic terminology tend to focus less on describing the specific actions in detail, relying on terminology alone to reveal the magical event.

It is also important to note that it is completely plausible that these terms were actually used prior to the fourteenth-century examples already described, but any evidence of earlier use has been lost over time. *Gerningar* can be found in other compound words within earlier examples of Old Norse literature, religious sermons and legal texts, prior to the known uses of weather-magic terminology listed in this section.[7] For example, the term *gerningamaðr*, referring to a man who has magical knowledge, can be found in multiple sources dating to the thirteenth century.[8] Furthermore, the use of *gerningar* in sources such as legal texts and religious sermons highlights that this type of magical action was not only limited to literary motifs that were used to enhance a narrative, it also existed within medieval Icelandic society as a social construct.

Járnsíða, a legal text from late thirteenth-century Iceland, includes a line that refers to 'þæir menn er geraz flugumenn[9] til at drepa menn þa er þæir eigo engar sakar við ... oc sua gørninga menn' (NGL 1: 265) [Those men who become men of flies [magic practitioners] to kill those men whom they have no charge against ... and also magical men]. Regardless of the fact that law texts were only written down following Iceland's conversion to Christianity and clearly carry a Christian bias, the inclusion of magical terminology in legal writings like this must suggest that these notions of magic were considered plausible in thirteenth-century Icelandic society. Although Scandinavia's adoption of Christianity and the continued acknowledgement, belief, or practice of magic are seemingly incompatible, Church authorities were unable to wipe such magical notions out of medieval Norse society (as with the rise of witch trials in the Early Modern period).

Galdrahríð

Galdrahríð, the final compound discussed here that is specific to weather magic, is formed from the Old Norse word *galdr*, plus *hríð*. Kirsten Hastrup (1990: 200) defines *galdr* as a term referring to a song, which de Vries (1977) notes is derived from the verb *gala*, meaning to sing or shout, or to recite magic formulae. Hastrup suggests that the linguistic emphasis of this term is an important key to understanding how this magic was vocally performed through songs or chants (Hastrup 1990: 200). These magical incantations would have been seen as vocal instruments to be used by the performer (Halvorsen 1960).[10] Mitchell (2019: 143) similarly notes that verse form and vocal expression, especially in the case of *galdralag*, signal performative magic, underscoring how changes in meter could mark moments of magical intervention in Old Norse texts.

Nicolas Meylan (2014: 35) suggests that by the thirteenth century in Iceland, the crucial element to successful *galdr* was less dependent on the singing/chanting aspect of the performance and more reliant on the content of what was uttered. However, Catharina Raudvere (2002: 95) notes that there are very few remaining medieval Scandinavian sources that contain an actual *galdr* text.[11] Regardless of whether *galdrar* were considered to be fictionalised or a reality during the time that these sagas were written down, Christian scribes may have been uncomfortable, or even prohibited by church authorities, to physically document these magical songs within their texts (Jóhanna Katrín Friðriksdóttir 2013: 50). Instead, the surviving sources only contain hints towards the importance of vocal incantations through terminology, such as *galdr*, and the depiction of magic performers chanting or singing to employ their magical skills (Raudvere 2002: 95). However, there are surviving examples of *galdralag*,[12] which is a poetic metre that scholarship suggests was intended to represent magic (Westcoat 2016). The actual word *galdralag* occurs very infrequently in the Old Norse corpus, and the only known medieval occurrence is found in *Háttatal*, which is preserved in the Codex Regius (GSK 2367 4°, c. 1300–1325). The connection between poetry and *galdr* will be discussed further in the third chapter on performance, which will further highlight the important link between poetry and using vocal spells to perform weather magic.

Meylan (2014: 36) suggests *galdr* terminology alone cannot provide us with the details of magic-working in medieval Scandinavia. However, in the case of weather magic, the examination of *galdr* terminology combined with the analysis of literary descriptions of weather magic found within a wide range of sources, allows us to piece together a better understanding of how this vocal magic was understood by contemporary society in medieval Scandinavia. The only existing example of the word *galdrahríð* is found within a version of the legendary *Hrólfs saga kraka* that is preserved in the manuscript AM 285 4°. The events in this saga are set in the late fifth and early sixth centuries and follow the adventures of Hrólfr kraki and his people, the Skjǫldungs. In its current form, the saga dates to the end of the Icelandic Middle Ages, however, the oldest of the 44 surviving manuscripts is from the seventeenth century (Byock 1998: viii).

The incident of weather magic occurs during the culmination of the saga, when Hrólfr goes to battle against his half-sister, Skuld, and her husband Hjǫrvarðr. During the confrontation, Skuld attempts to defeat Hrólfr's army by using her magical abilities to summon a deadly storm. This act is narrated by using the term *galdrahríð*, which implies that a vocal element is required for Skuld to successfully call the storm to battle: '[k]om nu áá su galldra hryd ad kapparner töku til ad falla huǫr vmm þueran annan' (Slay 1960: 123) ['such a [magical storm] now descended upon them that the champions began to fall, one across the other' (Byock 1998: 78)].

Although this is the only example of the compound term in use within the Old Norse corpus, there are many additional episodes that combine vocal charms and weather magic. Chapter 10 of *Fóstbrœðra saga*, for example, contains an example of vocal incantations altering the weather, although in this scenario the narrator chooses to use the term *fornu kvæði*, meaning 'old chants', rather than *galdr*: 'En Gríma hafðisk á landi uppi við um daginn ok minnisk á þau in fornu kvæði, er hon hafði í barnœsku sinni numit. Í því bili fellr andviðrit, þat er áðr hafði lengi á legit' (ÍF 6: 169) [Grima stayed ashore and recited some old chants she learned in her childhood. At that very moment, the wind that had been blowing inland for so long suddenly dropped (CSI 2: 353)]. Bernadine McCreesh (2018: 72) suggests that the author consciously decided to forgo the use of *galdr* in an attempt to distance Gríma's actions from ones closely tied to non-Christian practices. Similar to Dalla and Helga's actions describedin *Finnboga saga ramma*, Gríma's choice to use magic is made in an attempt to prevent vengeful bloodshed, a motive which Christian scribes and audiences might have interpreted as justifiable.

The inclusion of magic-related personal names can also highlight the connection between weather magic and vocal instrumentation. In *Eyrbyggja saga*, Þórðr arranges for Þórgríma galdrakinn, or Þórgríma magic-cheek (CSI 5: 181), to cast a storm against Bjorn during his travels across the heath (ÍF 4: 109). In the text, the author omits details surrounding the actions or instruments used by Þórgríma to create the magical storm. However, providing her with the nickname *galdrakinn* perhaps suggests that she relies on a vocal element when practising magic.

Following the previously mentioned episode in *Vatnsdœla saga* where Þórarinn challenges Starri to a duel, the text indicates that vocal elements were a requirement to magically combat the sorcerous storm. Once the storm is attributed to magical intervention, Bárðr stirfinn is asked to calm the magical tempest. Everyone in the scene joins hands to form a circle, which then moves three times in the opposite direction from the sun. Bárðr speaks in Irish and waves a cloth towards the mountain, while the group chants their approval: 'Hann bað þá handkrækjask ok gera hring; síðan gekk hann andsælis þrysvar ok mælti írisku; hann hað þá já við kveða' (ÍF 8: 96) ['He asked them to join hands and make a circle; he then went round backwards three times, spoke in Irish, and bade them all say "yes" outloud' (CSI 4: 65).]. The group performance clears the dreadful storm. While the term *galdr* is not directly used in this example, the action portrayed implies the crucial need for a verbal element, from both Bárðr and the group, for the ritual to produce a successful outcome.

Vernacular Remnants of Weather-Magic Terminology

While the terminology discussed above is drawn from medieval Old Norse literary and legal sources, there is evidence that concepts associated with weather magic endured in the vernacular traditions of the North Sea region well into the Early Modern period. Shetland Norn, the now-extinct Norse dialect once spoken in Shetland, preserves a handful of evocative weather-related terms that may descend from Old Norse magical vocabulary. Among these are *gandaguster*, meaning a sudden or violent gust of wind, and *ganfer*, a word for a crackling or atmospheric disturbance (Mitchell 2001: 68–69). Both appear to derive from *gandr*, an Old Norse term for magic, and relate conceptually to *gandreið*, the 'witches' ride' or magical journey (Mitchell 2001: 68–69). Although these terms were recorded centuries after the medieval Old Norse sagas, they offer compelling linguistic evidence of the long cultural memory of weather-related magical phenomena in the North Atlantic world.

These Shetland terms may not preserve a direct understanding of medieval magical practice, but they reflect an enduring belief in the possibility of magical manipulation of the weather, and especially the role of sudden, unsettling atmospheric events as signs of magical activity. Their survival underscores the potency and longevity of weather-magic motifs in vernacular consciousness, even after the Christianisation and legal suppression of such practices.

Seiðr

For a different perspective on the scene where Bárdr calms a magical storm, Einar Ólafur Sveinsson (ÍF 8: 96) compares the actions in *Vatnsdœla saga* to the famous divination scene in *Eiríks saga rauða*. He suggests that the linking of arms and chanting in *Vatnsdœla saga* is reminiscent of the *hringr* (circle) used around the *vǫlva*, Þorbjǫrg lítilvǫlva, when she employs *seiðr* as a prophetic ritual. If Bárdr's counter-magic is indeed a *seiðr* performance, it is not the only example of weather magic that links vocal instrumentation with a *seiðr* ritual. In the previously discussed scene from *Laxdœla saga* (Ch. 35), Kotkell raises a *seiðhjallr*, or high platform, in preparation for the spell his family casts against Þórðr and his ships. The use of the *seiðhjallr* infers that the ritual performed is one associated with *seiðr* magic. Moreover, the magical action described also includes the verbal addition of *galdrar*. The addition of a vocal aspect reveals that even when weather magic can be interpreted to fit within a certain type of magical performance, such as *seiðr*, the use of a vocal element is central to the spell's success.

In a very general perspective, Catharina Raudvere defines *seiðr* as a magical technique used for 'gaining knowledge about the future or trying to change the options for events to come' (Raudvere 2002: 111). In 1935, Dag Strömbäck conducted an in-depth investigation into the Old Norse notion of *seiðr*, which has since received an increase in academic focus over recent years (e.g. Dillmann 2006; DuBois 1999; Price 2002, 2019; Raudvere 2002; Heide 2006; Tolley 2009; Mitchell 2011; Meylan 2014). The exact etymology of *seiðr* has been disputed; however, several scholars such as Jan de Vries (1977), Ursula Dronke (1997: 133), and Eldar Heide (2006) have suggested the term is derived from a group of Indo-European words meaning 'to bind' in a magical context. Meylan (2014: 39) debates their proposed origin of *seiðr* by stating that he finds the link to binding one too good to be true, as the notion of binding was often used by twentieth-century scholars to define magic. Dillmann (2007: 859) proposes that *seiðr* is linked to a Greek term meaning 'song', which could suggest a similar link to vocal instrumentation as suggested by Raudvere (2002).

Episodes featuring *seiðr* performances are portrayed within a multitude of Old Norse sources, and the performers of this type of magic were often designated by using terms such as *seiðmaðr*, for male practitioners, or *seiðkona*, for female practitioners. Many occurrences of weather magic within Old Norse saga literature are tied to *seiðr*, and these performances will be closely analysed throughout many chapters of this study.

Conclusions

In conclusion, the examination of general magic and weather-magic terminology within Old Norse sagas reveals that the concept of weather magic was deeply ingrained in medieval Scandinavian society. Through the analysis of the terminology that is specific to acts of weather magic, *gerningaveðr*, *gerningahríð* and *galdrahríð*, alongside the contextual evidence provided by the scenes that employ these terms, this chapter reveals several pivotal findings that illuminate the role and contemporary cultural understanding of weather magic in Old Norse saga literature.

First and foremost, weather magic emerges as a potent narrative tool, offering characters a means of agency and exerting profound influence over the course of the plot. A clear illustration of this narrative function is found in *Vatnsdœla saga*, where Helga at Borg wields weather magic to avert a duel between her husband, Bergr, and Jǫkull. By conjuring an unnatural snowstorm, Helga disrupts the expected outcome of the conflict, asserting her control and reshaping the trajectory of the narrative. This narrative agency given to characters through weather magic extends beyond mere plot advancement, affording female characters like Helga a compelling opportunity for navigating and challenging the prevailing norms of a male-dominated society.

Central to understanding the nuances of weather magic in Old Norse saga literature is the precise terminology employed to denote instances of magical weather manipulation. Terms such as *gerningaveðr*, *gerningahríð*, and *galdrahríð* serve as linguistic markers, delineating the distinctive practice of weather magic within the broader field of magical practices. However, the scarcity of detailed descriptions of weather-magic performances underscores the reliance on contextual clues and terminology to convey weather-magic events. For instance, in *Hrólfs saga kraka*, the term *galdrahríð* refers to Skuld's invocation of a violent storm, accentuating the key role of vocal elements in the magical ritual.

Moreover, vocal incantations emerge as a recurring motif in weather magic, further underscoring the significance of verbal components in magical practices. This is exemplified in *Fóstbrœðra saga*, where Gríma's recitation of *fornu kvæði*, or 'old chants' coincides with a sudden end of the wind, highlighting the effectiveness of vocal instrumentation in manipulating the weather. The association between weather magic and vocal incantations suggests a performative dimension to magical rituals, wherein chanting or singing plays an important role in harnessing magical power and altering the weather to suit the needs of the practitioner. The connection between weather magic and vocal performances will be further demonstrated in the following chapter on weather-magic performances (Chapter 3).

The cultural importance of weather magic is further demonstrated through its integration into legal texts and religious sermons, reflecting widespread belief in these magical practices within medieval Icelandic society. As Mitchell (2019: 145) observes, the consistent embedding of magical vocabulary across legal, ecclesiastical, and literary sources highlights the social reality of magic and the deeply rooted conceptual frameworks that supported it in medieval Scandinavia. The portrayal of weather magic as a social construct underscores broader cultural attitudes towards magic and magical performances, and demonstrates its integration into everyday life. The overlap between weather magic and *seiðr*, a broader magical technique encompassing divination and manipulation of the future, emphasises the multifaceted and interconnected nature of magical practices in medieval Scandinavia. This connection is evident in *Eyrbyggja saga*, where Þórðr arranges for Þórgríma galdrakinn to cast a storm against Bjǫrn, once again highlighting the central role of vocal elements in magical workings and their cultural significance.

In conclusion, the study of weather magic and general magic terminology in Old Norse sagas offers invaluable insights into the intricate relationship between narrative, cultural beliefs, and how magical practices were perceived in medieval Scandinavia. By examining specific examples of weather magic and the terminology used to denote these magical performances, this chapter has offered a deeper understanding of how weather-magic functions in Old Norse saga literature, shaping character dynamics, plot developments, and cultural attitudes towards the performance of magic. Due to the volume of weather-magic episodes and weather-magic terminology, the magical manipulation of the weather emerges as an important literary theme through which to explore the cultural attitude towards magic in general in medieval Scandinavian society, illuminating its complexities and offering a deeper understanding of the magical themes that existed in the cultural and literary landscape of the time.

Notes

1 A version of this saga also exists in the *Flateyjarbók* manuscript but there is no use of weather-magic terminology.

2 Óðinn's vast reputation for performing weather magic will be explored in Chapter 6 of this study.

3 Pseudo-Vatnshyrna is composed of three fragments from similar dates that share similar palaeographical characteristics: AM 445 b 4to, AM 445 c I 4to, and AM 564 a 4to. For further information on the scholarly arguments for considering these fragments as a single volume, see Karlsson 1970 and McKinnell 1970. For further reading on the problematic uses of the now-lost Vatnshyra as a source, see Jóhanna Katrín Friðriksdóttir 2021: 102–108.

4 The outcome and negativity associated with this weather-magic scene will be further discussed in Chapter 2, in the context of beneficial versus harmful weather magic.

5 *Apologiae*, a term coined by Marianne Kalinke (1985: 319), are small passages that were added to the manuscript by the scribe, where the writer would comment on the 'truth-value' of the narrative (O'Connor 2009: 363). These passages

normally appear in the prologues or epilogues of the narrative and are often used to defuse the audience's scepticism about magical elements found within the text.

6 This *apologia* has been at the centre of many recent discussions, particularly by Ralph O'Connor (2009) and Philip Lavender (2018).

7 *Gerningafullr* and *gerningamaðr*, for example.

8 *Elís saga ok Rósamundar* (c. 1270), *Járnsíða* (c. 1271–1281), *Barlaams saga ok Jósafats* (c. 1275), *Hómilíur: Homiliae & sermones* (c. 1290–1334).

9 Richard Cleasby and Gudbrandur Vigfusson propose that the term *flugumenn*, or men of flies, stems from the notion that magic practitioners were said to cast a spell on flies and then send them to kill their enemies (Cleasby and Gudbrandur Vigfússon 1874, s.v. 'flugumenn').

10 For a deeper analysis of the term *galdr*, see de Vries 1977, Halvorsen 1960, Hastrup 1990, Raudvere 2002, Price 2002 and 2019, Dillmann 2006 and Meylan 2014.

11 There is an Early-Modern Icelandic text called *Galdrabók* that contains charms, spells and recipes, but it dates to the very late sixteenth century (Mitchell 2015: 64). Although some medieval texts do refer to magical books, like the magical book that raises a storm in *Söguþáttur Jóns Halldórssonar biskups*, there are no actual surviving examples from the medieval period. This topic will be further analysed in Chapter 3 on weather-magic performance.

12 See Westcoat 2016 for an overview of existing scholarship surrounding *galdralag* metre, as well as a well-presented argument on how *galdralag* is inextricably linked to magic.

2

READING THE WINDS

Weather Divination and Efficatory Weather Magic

Many scholars continue to divide the practice of magic in a medieval Nordic context into two subcategories: harmful or black magic, and beneficial or white magic (see Strömbäck 1935, 2000: 142; Tolley 1995: 58; Raudvere 2002: 80; Merkelbach 2019: 127). In this method of taxonomy, magic is placed into a harmful or beneficial magic category based on the performer's intentions (Raudvere 2002: 80). However, this sort of binary categorisation is problematic for the field of magical studies as it assumes that magical acts can, and should, be understood as strictly acts of good versus evil. Furthermore, determining whether an act of magic is definitively an act of good or bad is problematic in that it relies on the interpretive framework of the observer. The second half of this chapter will explore this problematic division further. As an alternative method of categorising magic, Neil Price (2019: 186) divides Norse magic into the spheres of domestic magic and aggressive or violent magic. In these allocations, Price assigns the manipulation of weather to the domestic sphere of magic, which also includes additional functions such as divination, bestowing good or bad fortune, attracting animals and fish for hunting, healing the sick, causing mild harm, and finally, communicating with the dead, unseen worlds, and gods. For the purposes of this study, I find the allocation of weather magic to the domestic sphere also problematic in that it overlooks the many functions that weather magic holds. For example, weather magic is prominently depicted as a force in warfare, as seen in the *Jómsvíkinga saga* during the Battle of Hjǫrungavágr. Jarl Hákon, facing defeat against the Jomsvikings, sacrifices to the deity Þorgerðr Hǫlgabrúðr, who conjures a violent storm of hail and lightning, turning the tide of battle in his favor. This episode underscores the strategic use of weather magic in combat, aligning magical performances with political and

DOI: 10.4324/9781003746058-4

military power. Similarly, weather magic plays a role in political-religious conflicts, as seen in *Ólafs saga Tryggvasonar* (Ch. 80), where the non-Christian chieftain Rauðr inn rammi attempts to thwart King Óláfr's fleet with a magical storm, only for Bishop Sigurðr to dispel it using Christian prayers and holy water. A similar struggle appears in *Ólafs saga helga* (Ch. 9), where resistance to Christian rule is met with weather manipulation, portraying the control of the elements as a contested source of power between religious and political factions. These examples, all of which will be further examined later in this study, illustrate that weather magic was far from a purely domestic practice; it was wielded in high-stakes conflicts to shape the outcomes of battles and ideological struggles.

In his comprehensive study of magic within medieval Icelandic literature, François-Xavier Dillmann (2006) divides his study of magic into the subcategories of 'divination' and 'magic that results in a physical change'. His research on physical magic is further separated into two sections: *fjǫlkynngi*, or the performance of magic, and those who perform *fjǫlkynngi*. Dillmann's study of magic opens with the examination of the divination scenes found in medieval Icelandic literature, and then turns to focus on the literary portrayals of *fjǫlkynngi*, which Dillmann categorises into the three following groupings: magic relating to warfare and weapons, magic that connects with the body and the soul, and finally, magic that deals with the four elements of nature. Although I find this type of definitive categorisation unnecessary in the field of magic due to the inevitable way in which certain magical acts overlap into multiple categories, I do acknowledge that focusing magic into smaller groups makes the large volume of material more digestible. The present study will therefore focus on Dillmann's third category of magic, which utilises the four elements, with particular emphasis on manipulating the weather. While there is no explicit evidence that medieval Scandinavians conceptualised weather magic through a structured four-element framework akin to Aristotelian or alchemical traditions, the manipulation of natural forces – air, water, fire, and earth – is deeply embedded in Old Norse cosmology and magical practice. Old Norse saga literature and other medieval sources from Scandinavia frequently depict weather magic as the control of wind, storms, and waves, suggesting an inherent connection between magical practice and the elements. For example, *Jómsvíkinga saga* describes Jarl Hákon invoking a violent storm through Þorgerðr Hǫlgabrúðr, illustrating a magical command over air and water, while *Gísla saga Súrssonar* portrays an old woman summoning an avalanche, invoking earth as an instrument of magical retribution. Similarly, *Hávamál* includes references to spells that calm the sea, reinforcing the idea that elements could be manipulated through incantation. These examples indicate that while a formal four-element theory may not have been an emic framework, elemental forces were nevertheless integral to medieval Scandinavian conceptions of magic, particularly in the realm of weather manipulation.

This study follows Clive Tolley's use of the term *efficatory magic*, which, while not recorded in the *Oxford English Dictionary*, has been employed in Old Norse magic studies to describe magic intended to bring about a tangible effect (Tolley 1995). Unlike *efficacious* magic, which primarily denotes effectiveness, the term *efficatory* emphasises the performative aspect of magic, or magic done with the explicit goal of directly influencing reality. This distinction is particularly useful in examining Old Norse sources, where spells, incantations, and ritual acts are described primarily in terms of their intended function rather than merely their success or failure.

Rather than using the emic term *fjǫlkynngi* to describe acts of magic that result in physical change, I adopt efficatory magic due to the lack of a singular Old Norse term that consistently refers to magical practice, particularly the performance of weather magic, as outlined in the previous chapter on terminology. Additionally, *fjǫlkynngi* carries a pejorative connotation in Old Norse texts and is never used neutrally to describe magic (Raudvere 2002: 88). By contrast, efficatory magic remains neutral and descriptive, avoiding the negative bias inherent in medieval terminology. Moreover, the use of this term instead of a strictly Old Norse one helps connect this study to the broader field of magical studies, providing a more precise and comparative framework for analysing weather magic across cultures.

Similar to Tolley's (1995: 58) research on *seiðr*, Dillmann's study separates divination from efficatory magic. Efficatory weather magic and divining the weather are portrayed as two very different means of magic in the Old Norse corpus. Therefore, I will apply the same categorisation in the next section by opening with an analysis of weather divination scenes, followed by an analysis of efficatory weather magic in Old Norse literature. The first half of this chapter will explore weather divination and what the literary depictions of these divination performances can reveal about the medieval Icelandic understanding of weather foretelling. The second half will then turn to the analysis of efficatory weather magic as it is portrayed throughout Old Norse literature, and the problematic division of such magic into harmful and beneficial magic categories.

Weather Divination

The Old Norse corpus reflects a Scandinavian interest in destiny and fate, and the desire to have some control over the future. Pre-Christian mythology reveals that both humans and non-Christian gods alike were thought to consult magic practitioners known for their skills in divining the future. In the eddic poem *Vǫluspá*, one of the most well-known examples of this motif of consulting divination specialists, Óðinn seeks the wisdom of a *vǫlva*, or prophetess skilled in magic, and learns about Ragnarǫk. Similarly, an interest in knowing what will happen in the future is also expressed when it comes to the everyday necessities of life. Knowing short-term and long-term weather forecasts falls under this category, as the success of daily activities, like farming or fishing, often relies on good weather. In a wider European context, weather and harvest forecasting is, and has always been, one of the most necessary forms of predictions (Cameron 2010: 66). This universal need to predict the weather is reflected in Old Norse saga literature, through scenes of characters consulting magic practitioners who have reputations for foretelling the future. In an Old Norse context, divination is often portrayed as an act of *seiðr*. Raudvere (2002: 120) notes that when used in Old Norse literature, divination is always part of a narrative strategy that is constructed in the text. These *seiðr* rituals can be conducted for the benefit of the performer but more often than not, they are used in text as a way for the practitioner to assist a recipient in need of support. As an example, the fourth chapter of *Eiríks saga rauða* (ÍF IV; CSI 1) contains the most comprehensive account of a *seiðr* divination ceremony found in the Old Norse corpus. Here, Þorbjǫrg *lítil- vǫlva* (little-seeress) is called to a farmstead because the community is experiencing a harsh winter and is seeking guidance. Her foretelling addresses the initial problem of the community wanting to know if the weather will improve their situation, and then fulfils the narrative function of outlining the storyline ahead, which is often the purpose of divination and dream motifs in the Old Norse sagas (Conroy 1980: 119; Raudvere 2002: 126). Þorbjǫrg's divination scene reveals that not all of the content included in the predictions portrayed in Old Norse saga literature is essential to the plot. In the case of foretelling the weather, the actual forecast divinations rarely have any influence on the narrative's storyline after the prophecy has been made. Rather, it is the foretellings included alongside or after the weather predictions that have any major effect on the story's plot line. In this divination scene from *Eiríks saga rauða*, a foretelling about Guðríðr's future is the only part of the seeress's prophecy, which subsequent parts of the story show to come true; the weather predictions are dropped as soon as they are made, as if they are of secondary interest. There are no examples of weather predictions alone that directly result in the divination recipients adjusting their actions and changing the narrative.

Although it is often overshadowed by the highly detailed description of her outfit and aspects of a *seiðr* ritual, Þorbjǫrg does indeed predict the weather during her divination ceremony in *Eiríks saga rauða*. The chapter begins by detailing a harsh famine in Herjólfsnes. As the most important member of the community, Þorkell is responsible for inviting a *spákona* (a seeress) into his farmhouse to conduct a *seiðr* performance, where the ritual performer will predict the harvest season to come. During her performance, Þorbjǫrg foretells that the bad harvests will shortly come to an end. Although her forecast prediction is not the most important aspect of her divination in terms of revealing the ensuing narrative, it is what initiated her visit to Herjólfsnes in the first place and therefore plays an essential role in developing the narrative's storyline. This example supports Raudvere's suggestion that divination scenes in Old Norse sagas are used as a narrative strategy to develop the plotline. Yet the decision to specifically use weather predictions as a motive to invite Þorbjǫrg to the farmhouse may suggest that weather divinations were considered a legitimate reason to host a *seiðr* ritual for the community.

A comparable example of weather divination is found in *Ǫrvar-Odds saga* (Ch. 2). Similarly to the *seiðr* performance in *Eiríks saga rauða*, the weather-divination motif is employed in *Ǫrvar-Odds saga* as a means to introduce a prophecy that will eventually cause the main character to move along the narrative's plot. The prophecy chapter in *Ǫrvar-Odds saga*, first included in the saga's oldest-known recension from c. 1300 (AM 343, 4°), opens with the introduction of Heiðr,[1] a seeress described as a '*vǫlva ok seiðkona ok vissi fyrir úorðna hluti af fróðleik sínum*' (Boer 1888: 11) [a *vǫlva* and *seiðkona* 'who had second sight, so with her uncanny knowledge she knew all about things before they happened' (Hermann Pálsson and Edwards 1985: 28)]. The text reveals that Heiðr's divination performances normally took place at feasts where 'sagði hon mǫnnum forlǫg sín ok vetrarfar eða aðra hluti' (Boer 1888: 11) ['[she told] people their destinies and forecasting the weather' (Hermann Pálsson and Edwards 1985: 28)].

Hearing that Heiðr is close to his farmstead, Ingjaldr orders his son Ásmundr to seek out and invite her to perform a *seiðr* ritual for their household. After delivering her prophecies to Ingjaldr, Ásmundr and the rest of the household, Heiðr then '*sagði ... um vetrarfar ok marga aðra þá hluti, er menn vissu eigi áðr*' (Boer 1888: 12) ['predicted the weather for the following winter and a lot more that was not previously known' (Hermann Pálsson and Edwards 1985: 29)]. This sentence describing Heiðr's prediction of the weather at Ingjaldr's farmstead is a later addition to the narrative, which is first found in the manuscript AM 471, 4° dated c. 1450–1500. Before departing from Ingjaldr's farmstead, Heiðr notices Ingjaldr's foster son, Oddr, who is sceptical of her performance and had not yet received a prophecy. It is here that Heiðr delivers her most essential prophecy in terms of revealing the subsequent plot, which predicts Oddr's upcoming adventures, as well as his ultimate demise. As the older manuscript does reference Heiðr's reputation for foretelling the weather prior to her visit to Ingjaldr's farm, it is unlikely that the addition of the weather prediction in the fifteenth-century manuscript reveals anything about shifting attitudes towards non-Christian magic. Instead, the weather divination is most likely a coincidental addition, given that this version of the saga tends to provide more details to add colour to the story.

It is interesting to note that the older recension includes the initial link between Heiðr's knowledge of *seiðr* and her ability to forecast the weather, but it does not actually demonstrate that connection by having her foretell the weather in the described *seiðr* ritual. Regardless, similar to the scene in *Eiríks saga rauða*, the addition of the weather divination in *Qrvar-Odds saga* has no real influence on the narrative's plot. Although the weather divination does not directly influence the plot of *Qrvar-Odds saga*, Raudvere's (2002: 120) suggestion that divination scenes in Old Norse sagas serve as a narrative strategy to develop the plotline remains valid. In the case of Heiðr's divination, while the weather prediction itself does not alter the narrative, it is the additional prophecies she provides, particularly concerning Oddr's future, that propel the plot forward. This demonstrates that, even when the weather divination appears inconsequential, divinatory practices still serve a significant role in revealing future events and shaping the progression of the narrative. Perhaps these seemingly irrelevant occurrences of predicting weather and future harvest outcomes suggests that weather prophecies were an expected or normalised element within the medieval Icelandic understanding of a *vǫlva*'s divination performance. The two mentions of weather forecasts that are described above play no role in what is considered to be the normal narrative function of divination, which is to reveal the upcoming storyline. Nor do these weather forecasts change the character's actions within the narrative. Therefore, their inclusion implies that they are incorporated into these divination scenes because weather forecasts are an element that was generally expected from a *seiðr* performance. It is understandable that weather divinations would have held some type of cultural importance in medieval life because so many key aspects of daily living were tied to the weather. Although it is tempting to try to tie weather divinations to Viking-Age practices, the existence of two examples alone from manuscripts written centuries after the Viking Age does not provide substantial evidence to make a claim that the weather divination portrayed in the sagas is a remnant of an older Scandinavian tradition. The only thing we can securely take from these divination scenes is that it is perhaps plausible that medieval Icelanders understood weather divination to be a normal part of the *seiðr* performance.

An additional function of weather divination worth mentioning, albeit also very rare in occurrence, is unusual weather occurrences as a means of foreshadowing events. One literary example of this type of foreshadowing is found in *Eyrbyggja saga*. In Chapter 11, a mysterious cloud suddenly appears and starts to rain blood. The blood quickly dries everywhere except on the hay, which Þorgunna had previously spread. Þorgunna interprets this event to mean that someone would likely die. Unlike the previously described divination scenes in *Eiríks saga rauða* or *Qrvar-Odds saga*, where there are aspects of the divination performance that could influence either the individual or the a wider-community, Králová (2017: 32) suggests that omens in the family sagas are only relevant to the characters in focus and that the omens are closely bound to the timeframe of the saga narrative. This is reflective of the blood-rain omen in *Eyrbyggja saga*, as the blood-rain is not interpreted as a premonition of an event that would influence the whole community in a long-term scenario, like a prediction of a bad harvest that would cause a depletion of food sources. Instead, the premonition is tied to the near future, which is immediately realised following the omen, and is linked to the characters in the narrative foreground when the omen occurs. Once the interpretation of the weather omen has been made, Þorgunna falls ill and dies shortly thereafter. Before she dies, she warns Þóroddr that all of her personal belongings, including her fine bed linens, must be destroyed as soon as she passes. Instead of following Þorgunna's advice, Þóroddr's wife becomes enthralled with Þorgunna's possessions and convinces her husband not to destroy them. The disregard of Þorgunna's instructions leads to strange occurrences, including hauntings and unexplained illnesses, and eventually results in the death of Þóroddr and Þorgríma *galdrakinn*.[2] The strange occurrences end as soon as Kjartan destroys Þorgunna's belongings, and thus brings a close to the events tied to the weather omen.

A similar weather omen foreshadowing death is found in *Laxdœla saga*. In Chapter 76, a *vindsgnýr mikill* (ÍF 5: 222), or great gust of wind, against Þorsteinn's house is interpreted as a bad omen, foreshadowing that wind will be the death of Þorkell. Once again, by foreshadowing the death of Þorkell, the weather omen is linked to the specific timeframe of the saga narrative and only affects the characters relevant to the storyline in that moment. The day after the omen, Þorkell and his crew set sail. Once at sea, a squall catches the sail of the boat and causes their ship to capsize, causing Þorkell and his men to drown. The death of Þorkell mainly affects Þorsteinn, the character who originally interpreted the gust of wind as a bad omen. This example fits into Králová's aforementioned proposal that omens only affect the characters in focus and occur within the timeframe of the saga timeline. Perhaps the inclusion of the weather omens in *Eyrbyggja saga* and *Laxdœla saga* can be interpreted as a reflection of a medieval superstition surrounding extreme or seemingly out-of-nowhere weather occurrences, but again, the lack of additional incidences or variations of weather omens in the Old Norse saga literature makes it impossible to draw concrete conclusions.

Efficatory Weather Magic

Clive Tolley defines efficatory magic as 'effecting a physical change by means of magic' (Tolley 1995: 58), i.e. in the case of weather magic, using a spell to manipulate the weather. The previously discussed scene in Chapter 35 of *Laxdœla saga*, where Kotkell and his family summon a storm via *seiðr*, is a clear example of efficatory magic. In this case, the family's magical action brings a physical change to the weather, which results in the capsizing and destruction of Þórðr's ships. Similarly to the way in which scholars divide magic into different categories, existing scholarship (Tolley 1995: 58) subdivides efficatory magic into two categories: magic that is used for beneficial purposes, and the seemingly more common category of harmful magic. While the distinction between 'good' and 'bad' magic may seem straightforward, it is a classification widely employed in current scholarship with little critical reflection on its implications. By accepting these categories at face value without examining their historical and cultural contexts, scholars risk perpetuating medieval biases, particularly the demonisation of non-Christian and minority cultures. Many of the sources that inform our understanding of medieval magic were written from elite, Christian perspectives that positioned certain magical practices as inherently harmful or malevolent, often as part of broader efforts to marginalise specific groups, such as the Saami or other Indigenous populations. Without a thorough investigation of how each episode of magic is framed, we inadvertently reproduce these medieval narratives rather than critically analysing them. Thus, rather than simply reapplying a 'good versus bad' framework, this study seeks to unpack the motivations behind such classifications, considering how they were constructed in medieval sources and how they continue to shape modern interpretations of Old Norse magic.

On the surface, this division between 'good' and 'bad' magic appears to work when investigating portrayals of efficatory weather magic in the Old Norse corpus. Although weather magic is most often employed in Old Norse literature as a means to summon sudden blizzards and deadly sea storms, which is seen as a very disruptive magic (Merkelbach 2019: 127), there are also literary examples of magic performers seemingly manipulating the weather for beneficial purposes. Bárðr's efforts to counteract the magical storm that is summoned in *Vatnsdœla saga* (Ch. 34) is an illustration of a weather-magic performance which stops an unnatural storm, creating a solution for the overall advantage of the community.

Nevertheless, categorising weather magic, or magic in general, into either beneficial or harmful classifications is not as straightforward as determining whether the practitioner has virtuous or malicious motives. Furthermore, there is no surviving evidence that medieval Scandinavians themselves would have categorised magic in this way. As highlighted in the introduction of this study, the medieval Icelandic law code *Grágás* divides magic into two categories of severity. The first, less severe form of magic is described: 'Ef maþr ferr með galldra eþa gørningar. eþa fiolkýngi. þa ferr hann með fiolkyngi ef hann queðr þat eþa kennir. eþa lætr queða. at ser eþa at fe sinv' (Grágás I: 22). [If someone uses spells or witchcraft or magic – he uses magic if he utters or teaches someone else or gets someone else to utter words of magic over himself or his property (Dennis et al. 2006: 39)]. The second, more severe form of magic is outlined: 'Ef maþr ferr með fordæs skap. þat varþar scogGang. þat ero fe eþa mavnnvm' (Grágás I: 22). [If a man practices [a more serious form of magic], the penalty for that is full outlawry. It is [a more serious form of magic] if through his words or his magic a man brings about the sickness or death of livestock or people. (Dennis et al. 2006: 39).] Rather than separating magic as harmful and beneficial magic, this section in *Grágás* suggests that from a legal viewpoint, all magic was interpreted as unlawful and problematic in medieval Iceland.

When discussing the division of medieval Scandinavian magic into harmful or beneficial categories in modern scholarship, Raudvere (2002: 80) explains that magical practice in itself should not be divided into these two categories. Rather, it is the purpose for which the magic is performed that should determine whether it is classified as beneficial or harmful magic. Following this line of argument, using magic that results in someone's death cannot automatically be classified as harmful, or black magic. Instead, Raudvere is suggesting that scholars should examine whether the performer's motives stem from a virtuous or malicious attitude. Although seemingly straightforward, relying on the performer's intentions alone to determine whether the use of weather magic, or magic in general, is either beneficial or harmful can be problematic in many cases. In the remainder of this section, I will use the portrayal of weather magic in Old Norse literature as a case study to demonstrate that magic cannot simply be categorised into the binary classifications of beneficial and harmful, good and evil, or black and white magic. There are many scenes depicting weather magic where depending on whose perspective is examined (the performer's, the recipient's, or the narrator's), the event could potentially fit into either a harmful or beneficial magic category. Furthermore, the purpose for which the magical act is conducted may differ from what contemporary society considers to be beneficial or harmful, particularly when the performance is depicted via a literary source. Rather than solely relying on the performer's intentions, the audience and scholars must also consider the narrative tone to determine in which category the magical act belongs, if it belongs in a category at all. Often the difference between understanding a magical event as harmful or beneficial lies within the interpretive framework of the observer, or the one who is narrating the occurrence. Focusing on the performer's motives allows us to assess why the magical act was committed, while the surrounding narrative should reflect some aspect of the contemporary society's attitude towards the scenario in which the magic is performed. Assessing an act of magic from multiple perspectives reveals that the use of magic and the perception of its use is too complicated to be categorised in a simple binary system.

The motif of employing weather magic in a political, Christian versus non-Christian altercation highlights how tricky it can be to determine whether the use of magic is beneficial or harmful without an analysis of both the performer's intentions and the narrative tone.[3] Often in these situations, the performer's use of weather magic stems from the seemingly noble intentions, from an Icelandic perspective, of resisting colonisation and taxation, maintaining their religious traditions, and preserving their society's way of life. Nevertheless, in the majority of these cases, the narrator's tone reveals a clearly Christian and monarchist bias towards the exchange. In these episodes, the text almost always portrays the performance of weather magic as an adverse tool used to resist the justified attempts of a Christian king or emissary to convert, tax, or colonise a group of non-Christian individuals. The *konungasǫgur*, for example, frequently portray encounters between Christians and non-Christians through a narrative framework that ultimately legitimises the Christian and royal cause, even when the resistance of non-Christian figures is presented with a degree of sympathy or admiration (Harris 2008: 240–244).

A case exemplifying this Christian and monarchist bias is found in a scene from *Óláfs saga helga* (Ch. 9), described in the previous chapter on terminology, where an Indigenous community summons a storm to fight off King Óláfr's attempts to colonise their region. Although the Indigenous intentions of using weather magic can be construed as beneficial from their own perspective, the tone of the surrounding narrative clearly favours the king's Christian and political efforts. King Óláfr is portrayed as having the ability to overcome the magical storm through the king's good fortune (*Heimskringla* II: 8), which is portrayed as having more power than the Indigenous magic. This attitude in the narrative makes it apparent that King Óláfr is portrayed as the Christian main character of this story, while the Indigenous community is represented as the non-Christian obstacle. This example highlights how Christian 'magic', or good fortune, is often used throughout medieval Icelandic literature as a tool to conquer non-Christian societies. In the *konungasǫgur* in particular, weather magic is often outmanoeuvred by Christianity. This type of Christian 'magic', or power to combat non-Christian magic, is normally presented in the text as justified or morally superior to non-Christian magic. However, to the general audience, the Christian motives can be viewed as similar to the motives of the non-Christian magic practitioners.

As previously noted, it is crucial to remember that the narrator's tone is normally a reflection of the contemporary society's mood or attitude towards the topic of discussion. Old Norse saga literature was composed by Christian redactors that are, for the most part, writing in a colonial society ruled or heavily influenced by a Scandinavian king, in a period where Christianity was the dominant religion. As a result, the narrative tone of these stories reflects a colonised, Christian society and it is therefore unreliable to use their narrative tone when determining non-Christian Scandinavian attitudes towards certain magical practices from the past. Sirpa Aalto (2010: 184) suggests that in the case of *Heimskringla*, there are no instances where the use of magic is described in a positive manner. However, as shown through the examples above, determining the positive value of a magical event is not so straightforward. The inferences that are made when using the narrative tone to determine whether an act of magic is seen as justifiable or inexcusable can only be applied to the redactor's aims or agenda, and possibly to the text's contemporary audiences. There is no way to definitely designate these acts of magic into binary categories of harmful or beneficial because there are too many contradictory elements to consider, depending on who is performing the magical act and who is interpreting it.

Some Old Norse sagas contain descriptions surrounding the use of weather magic to resist religious conversion, taxation, and colonisation that are more ambiguous than the scene in *Óláfs saga helga*. *Færeyinga saga*, for example, holds one of the more subtle depictions of colonisation and religious conversion in the Old Norse corpus (Harlan-Haughey 2015: 375). In *Færeyinga saga*, the weather-magic motif is used as a means for Þrándr, a Faroese magic practitioner, to ward off a Norwegian plan to bring the Faroe Islands under Norwegian control and convert the Faroese people to Christianity. When Sigmundr, a Faroese man commissioned by King Óláfr to convert his own people to Christianity and collect taxes from the island, tries to approach the islands to find Þrándr, his crew is put off course by strong currents and winds:

> Nú er at segja frá þeim Sigmundi at byrr kemr á fyrir þeim ok sigla nú at eyjunum, ok sjá þá at þeir eru komnir austan at Eyjum, ok eru þeir menn á með Sigmundi at kenna landsleg, ok eru þeir mjǫk komnir at Austrey. Sigmundr sagði at hann mundi þat helzt kjósa at fá vald á Þrándi. Ok er þá berr at eyjunni kemr á mót þeim bæði straumr ok stormr, svá at ekki er nálægt um at þeir næði eyjunni; fá tekit í Svíney, með því at menn váru kœnir ok liðgóðir.
>
> *(ÍF 25: 51–52)*

> Now it is to be told about Sigmundr's party, that a fair wind rises for them and they now sail for the islands and see then that they have approached the islands from the east, and there are men on board with Sigmundr that know the landscape, and they are nearly come to Austrey. Sigmundr said that what he would most like was to get his hands on Þrándr. And when they were carried to the island there comes against them both current and gale so that they got nowhere near reaching the island; they managed to land on Svíney, since the men were skilful and capable.
>
> *(Faulkes 2016: 30)*

At this point in the narrative, there are no textual indications that magical intervention is causing the poor weather that is hindering Sigmundr's voyage. Without the use of any magical terminology, the weather blocking Sigmundr and his crew could simply be a result of natural strong current and heavy winds, however unlucky. Yet as the saga progresses (Ch. 31), it becomes clear that the weather thwarting Sigmundr's success in reaching Þrándr's location on the Faroe Islands is a magical intervention on behalf of Þrándr himself:

> En er Þrándr verðr þess varr at Sigmundr ætlar at flytja hann á konungs fund, þá baðsk hann undan þeiri ferð. En Sigmundr lét þat ekki tjá, ok slógu landfestum þegar byr gaf. En er þeir váru eigi langt í haf komnir, þá hittu þeir bæði í strauma ok storm mikinn; urðu við þat aptrreka til Færeyja ok brutu skip í spán ok týndu fé ǫllu, en monnum varð borgit flestum. Sigmundr barg Þrándi ok mǫrgum ǫðrum. Þrándr sagði at eigi mundi þeim ferðin takask slétt ef þeir léti hann nauðgan fara. Sigmundr sagði at hann skyldi fara allt at einu, þó at honum þœtti illt. Tók Sigmundr þá skip annat ok fé sitt at fœra konungi fyrir skattinn, því at Sigmund skorti eigi lausafé. Láta þeir í haf í annat sinn; komask nú lengra áleiðis en fyrr; fá þó enn mótviðri stór ok rekr þá aptr til Færeyja ok lesta skipit. Sigmundr sagði at honum þótti mikit farbann á liggja. Þrándr sagði at svá mundi fara hversu opt sem þeir leitaði til, svá at þeir flytti hann nauðgan með sér.
>
> *(ÍF 25: 73–74)*

> But when Þrándr realises that Sigmundr is planning to take him to see the king, then he asked to be excused from this journey. But Sigmundr was having none of it, and loosed the moorings as soon as there was a favourable wind. But when they had got not far over the sea, then they met both currents and a great storm; as a result they were driven back to the Faroes and broke the ship to pieces and lost all the money, though most of the men were saved. Sigmundr saved Þrándr and many others. Þrándr said that the voyage would not go smoothly for them if they made him go under compulsion. Sigmundr said that he should go all the same, even if he did dislike it. Sigmundr then got another ship and his own money to take to the king for the taxes, for Sigmundr had no lack of ready money. They set out to sea a second time; they now get further on their way than before; yet again they meet strong contrary winds and are driven back to the Faroes and their ship is damaged. Sigmundr said that he thought there must be some powerful obstacle to their going. Þrándr said that it would go the same however often they tried, as long as they made him go with them under compulsion.
>
> *(Faulkes 2016: 41)*

In this segment, Þrándr repeatedly warns Sigmundr and his men that as long as the crew tries to force Þrándr to go on the voyage against his will, the weather will continue to push against them. Throughout the saga, the portrayal of Þrándr's magical resistance is more ambiguous than what is normally represented in anti-political subjugation and anti-conversion magic found in the *konungasǫgur*. Rather than exercising an unmistakably negative representation of the pre-Christian, uncolonised agenda, *Færeyinga saga* portrays Þrándr's actions of resisting conversion and colonisation in a way where Þrándr's perspective and motives seem understandable, and possibly justifiable, to the reader (Harlan-Haughey 2015: 351–352). Although Þrándr's scheming actions are sometimes depicted to the point of being comical, they are always calculated and never executed without clear personal reason (Johnston 1994: ii). Regardless of how Þrandr's magic is portrayed, we can largely interpret Sigmundr's survival through the many occurrences of weather magic and the overall success of his efforts as a clear indication that his assignment of converting and taxing the Faroe Islands was seen as potentially justified by the saga's narrator. However, Sigmundr's actions are also portrayed in a similar manner. In this saga, both characters' actions are presented as justified from their own perspectives. As a result, the weather magic depicted throughout the narrative can neither be categorised as objectively harmful or beneficial. Once again, a binary categorisation of weather magic into harmful or beneficial is neither helpful nor necessary.

Harmful Weather Magic?

There are, however, cases of weather-magic performances where both the narrative tone and performer's intentions appear unmistakably to indicate that the magical act in question should be seen as either harmful or beneficial. The episode in *Laxdœla saga* (Ch. 35) where Þórðr and his crew drown as a result of Kotkell's magical tempest, is an example that, on the surface, could easily be interpreted as harmful weather magic. Kotkell's intentions of bringing the storm are retaliatory in nature and the magic is performed to cause the deliberate outcome of harming Þórdr. However, from Kotkell's perspective, his intentions for performing magic are positive in that he is reacting against the detrimental claim that Þórdr made against him. The surrounding narration echoes the hostility of the performer's intentions by using magic terminology like *mjǫk fjǫlkunnigr*, *galdrar* and *seiðmenn*, all of which are pejorative terms. The terms *seiðmaðr* and *seiðkona* were considered to be insults in medieval Iceland and using the words in Old Norse literature always denotes disgraceful individuals (Simek 1993: 200).[4] The harmfulness of Kotkell's magical act is emphasised by the death of Þórdr and his crew. That being said, the result of the magic achieved what Kotkell and his family intended, and they themselves would not see their acts as unjustified. Although it seems to be solely harmful on the surface, this episode can neither be clearly categorised as bad or beneficial due to differing perspectives.

Similarly, Auðbjǫrg's motive for her use of weather magic in *Gísla saga Súrssonar* (Ch. 18) is retaliatory in nature, which could easily be interpreted as a harmful motive. That being said, seeking revenge in medieval Iceland is often seen as necessary, and it is because her son, Þorsteinn, is wounded in the first place that she seeks revenge by sending a snowstorm and avalanche to Bergr's farmhouse. From Auðbjǫrg's perspective, she is simply a mother protecting and avenging her son. However, the events that immediately follow Bergr's death demonstrate how the narrative reflects a sense of unfairness surrounding Auðbjǫrg's weather-magic use. As a means of punishment and revenge for the death of Bergr and his household, Auðbjǫrg is taken to Saltnes and stoned to death. Þorsteinn seeks temporary safety at Gísli's farmstead and then flees Iceland. His decision to escape the country implies that Þorsteinn himself could have been held accountable for his mother's deadly use of magic.

Auðbjǫrg's death and Þorsteinn's flight indicate that Auðbǫrg's use of magic to seek revenge was considered unfair. By using magic, Auðbjǫrg avenges her son in a manner outside of the realm of acceptable methods for seeking revenge. Additionally, her use of magic prevented the community, and the victims themselves, from knowing who caused Bergr's and the others' deaths. William Ian Miller (1986: 102) explains that once a particular deed is perceived as wrong by a community, it is necessary to find and assign blame to the one who committed the misfortune. In the context of a medieval Icelandic blood feud, knowing the perpetrator of such a deed allows the community to know that a person has avenged their grievances. In this case, Auðbjǫrg's magic does not automatically allow for this. Her use of magic creates the problem that the perpetrator and the motives (i.e. to seek revenge) are initially unknown, and therefore the crime committed becomes the more serious crime of murder, rather than someone outwardly seeking revenge. Furthermore, her use of magic could also be construed as unfair by the surrounding community because Auðbjǫrg did not avenge Þorsteinn completely on her own, without a type of magical assistance. This idea that learning and using magic was considered unfair by medieval Icelandic society is reflected in the second chapter of *Bósa saga ok Herrauðs*. Here, Bósi explains to his foster mother that he does not want to learn magic from her because he does not want to have the reputation of having achieved anything through trickery, but rather on his own account.

The retaliatory weather-magic motif is not unique to the *Íslendingasaga* group of Old Norse sagas. In the eleventh chapter of *Þorsteins saga Víkingssonar*, a *fornaldarsaga*, weather magic is used to avenge the death of a brother. This saga portrays the families of Víkingr and Njǫrfi who are very competitive with each other. The Víkingssons play a dangerous game against the sons of Njǫrfi, which leads to one of the Víkingssons killing one of the sons of Njǫrfi. Víkingr sends the son who killed Njǫrfi's son to Lake Vänern, along with Þorsteinn and one other brother. Njǫrfi's eldest son, Jǫkull, asks Ógautan, a man known for his skills in magic, to help to exact revenge for the death of his brother. Ógautan agrees to help and uses his *veðrbelgr* ('weather-bag';

a magical bag that produces winds and foul weather) to conjure a frost that freezes Lake Vänern:

> 'Nú vil ek', segir Ógautan, 'segja þér, at ek á einn belg, þann er veðrbelgr heitir, enn ef ek hristi hann, þá stendr ór honum stormr ok vindr með svá stórri grimd ok kulda, at innan þriggja nátta skal lagðr svá sterkr ís á vatnit, at ríða skal mega hestum þótt vili'.
>
> *(FSN 2: 26)*

> 'Now I will tell', said Ógautan, 'that I have a bag called the weather-bag. If I shake it, storm and wind will blow out of it, together with such biting frost and cold that within three nights the lake shall be covered with such strong ice that you may cross it on horseback, if you wish'.
>
> *(own translation)*

This magical act allows Jǫkull and his brother to cross the lake and attack the Víkingssons. Later in the saga (Chapter 19), Jǫkull once again asks Ógautan to employ his skills in weather magic to retaliate against Þorsteinn. In this scene, Ógautan uses incantations rather than his magical weather-bag to summon several storms in the attempt to shipwreck Þorsteinn.

The retaliatory motive is one of two frequent motifs that Old Norse sagas depict as reason for performing what has previously been considered as harmful weather magic, the second motif being belligerent weather magic. Lyonel Perabo (2016: 125) defines belligerent weather magic as being offensive for no apparent reason. More often than not, this inclusion of the so-called belligerent weather-magic motif is found in the *fornaldarsǫgur*. To demonstrate belligerent weather magic, Perabo (2016: 125) refers to the following scene in *Ketils saga hængs*. In the third chapter of this saga, Ketill sets out for a fishing expedition close to winter, against the advice of his father, Hallbjǫrn. After he sets sail, Ketill's ship is caught in a violent gale that drives him away from the harbour and towards the coast of Finnmark. Once he is able to reach land, Ketill sets up camp and goes to sleep. During the night, Ketill is awakened by a *trǫllkona* shaking his ship, while the extreme winds are still blowing. Ketill leaps into the boat and rows away, fighting against the harsh gales. Eventually, Ketill and his boat are saved by a shapeshifter in the form of a whale and Ketill makes it to land and meets Brúni. When describing his plight to Brúni, Ketill attributes the weather magic to the Saami by delivering the following verse:

> Hér mun ek þiggja, hygge ek at valdi Finns fjölkýngi feikna veðri;
> ok í allan dag
> einn jós ek við þrjá, hvalr kyrði sjá,
> hús mun ek þiggja.
>
> (FSN 2: 117)

> Here I'll seek shelter! I suspect that a Finn wielded the windstorm
> with [magic] and spells. All day I bailed,
> doing three men's work. The whale soothed the sea.
> I'll shelter here.
>
> (Waggoner 2012: 10)

This episode in *Ketils saga hængs* gives no indication as to why the narrator chose to include the episode of weather magic, other than it plays a necessary role in moving the plotline, nor is the Saami motivation for summoning a storm to jeopardise Ketill's fishing expedition clear. The weather magic used by the Saami seemingly comes out of nowhere and the magical act is not explicitly depicted as retaliatory in nature. As it is implied that the magical act is intended to destroy the boat and bring harm to Ketill, it is understandable why this episode of weather magic has previously been interpreted as 'belligerent', and potentially belonging in a harmful magic category. However, this categorisation of the Saami's use of weather magic prevents the consideration of a marginalised point of view. Although their motivations are not clearly outlined in the episode above, the Saami most likely considered their magical actions as justified to some degree. The saga narrator is very explicit about Ketill's boat ending up very close to the shore of Finnmark, which could be seen as giving the Saami inhabitants a perfectly sensible reason to want to wreck his ship before it is able to reach land. Depictions of direct and violent attacks on Saami villages are prevalent throughout the *Hrafnistamannasögur*. Therefore, it is unsurprising that the Saami are unwelcoming towards an unknown ship approaching their village.

The altercation between the Saami and Ketill could also potentially be a reflection of contemporary tensions regarding fishing rights, and these tensions could be viewed as another justification for the Saami's use of weather magic. Evidence from the Middle Ages suggests that there was an established cooperation between the Norse and Saami when it came to the fishing industry. *Historia Norwegiæ* states that fish was one of the trading goods offered by the Saami to the Norse, and there is evidence that the coastal Saami were heavily involved in the dried codfish industry (Hansen and Olsen 2014: 154). Furthermore, multi-room houses found in Finnmark have been interpreted to suggest cooperation between Saami and Norse fishers (Hansen and Olsen 2014: 210). However, prior to Ketill embarking on the fishing expedition, the narrator explains that during the time that this story takes place, there was a great famine in Hálogaland and much of the people's livelihood had to come from the sea. Scientific research (Dybdahl 2012: 1163) confirms that harsh climates and famines were very much a reality in medieval Norway, including Hálogaland, which has been reflected throughout Old Norse literature.[5] Regardless of whether there was an established cooperation between Saami and Norse fishers, these famines would surely have resulted in competition surrounding diminishing food sources.

The lack of available food sources would explain any depictions of tension involving fishing rights, which could in turn serve as a motivation for the Saami to use weather magic against Ketill's attempts to fish off their coast.

Once again, placing the use of weather magic into the category of harmful magic, or applying the term 'belligerent' to the act, relies on an oversimplified analysis and does not consider the magical action from multiple perspectives. In many cases, like the Saami's use of magic against Ketill and the previously noted struggle between King Ólafr and the Saami in *Ólafs saga Helga* (Ch. 9), considering the use of magic as outright harmful ignores the perspective of voices that are normally marginalised in medieval Icelandic literature.

Beneficial Weather Magic?

As with harmful weather magic, there are cases where the performer's intentions and the narrative tone seem to align in indicating an act of magic as overall beneficial. However, these instances are very rarely portrayed in Old Norse literature. The previously considered episodes from *Vatnsdœla saga* (Ch.34) and *Finnboga saga* (Ch. 34), in which Helga or Dalla call a blizzard to prevent Bergr from participating in a duel, have both been described as illustrations of weather magic performed to create a beneficial outcome (McCreesh 2018: 69–70; 75). In each of these passages, magical intervention is a means to stop a duel from taking place and avoid an outcome where an individual of importance to the performer obtains injury or death. From the perspective of Helga or Dalla, summoning the blizzard is motivated in a desperate attempt to save Bergr, their partner, from injury. However, the narrative tone surrounding the episodes of weather magic in *Vatnsdœla saga* and *Finnboga saga* does not echo the seemingly helpful intentions. As noted in the previous chapter on weather-magic terminology, in the case of *Vatnsdœla saga*, the only language pointing to magical interference is the use of the term *gerningaveðr*:

> Svá er sagt, at þann sama morgin, at fara skyldi til hólmgǫngu þessarar, var komit á svá mikit kafafjúk með frosti, at engu vætta var út komanda. Þeir kváðu gørningaveðr verit hafa ok kenndu þat Helgu frá Borg.
>
> *(ÍF 8: 89–92)*

> It is said that the very morning on which they were to go to the duel, there was such a thick fall of snow and a frost that not a soul ventured out … [They] claimed that there had been [a magical storm] and blamed it on Helga at Borg.
>
> *(CSI 4: 45–46)*

As it is normally used in a negative way, the term *gerningar* clearly expresses a problematic use of magic. Furthermore, Helga is described as '*margkunnig um flesta hluti*' (ÍF 8: 89), or 'knowledgeable of magic', which also suggests a tone of disapproval. Although there are no further words that indicate magic, nor is the magical act described in detail, the text designates the performer and magical storm in a manner that implies hostility. The terminology used in this episode implies a general disapproval of magic, and therefore the actual act of weather magic probably carries the same sense of hostility from the narrator's perspective. In the *Finnboga saga* episode, Dalla herself takes credit for the magical storm, rather than the blizzard being attributed to her by others: 'Skal ek gera þat veðr at hvárki yðr né ǫðrum mǫnnum sé úti vært' (ÍF 4: 311) ['I shall bring about weather that neither you nor the others will care to be out in' (CSI 3: 254)]. Here, the narrative leaves no room for doubt as to who caused the change in weather, and that it was a result of direct magical interference. In *Vatnsdœla saga*, Helga's magical interference is only a rumour and therefore the audience may be more sympathetic towards Bergr's caution, as a real-life magical storm would be a reasonable excuse to avoid going outside. In *Finnboga saga*, however, it is clear that Bergr's failure to come to the duel was the result of magic, which confirms his cowardice to the audience. Although the language used in Dalla's declaration does not include magic terminology that would normally indicate a general sense of disapproval surrounding her use of magic, nor does the manner in which the storm is actually described stem from malicious motives,[6] the outcome of her magic reflects the unhelpfulness of her choice. Upon seeing the blizzard, Bergr chooses to stay at home and forgo the duel, while his opponent braves the storm. In both the *Vatnsdœla saga* and *Finnboga saga* episodes, Bergr is seen by others as a coward for not attending the duel. Taking into consideration the performers' motivation and how the narrator portrays the event in *Vatnsdœla saga*, it is clear that these two perspectives contradict each other. If the rumours are true that Helga is responsible for the storm, from her perspective, she performs the weather magic to save a loved one. In both portrayals of this scene, the magic causes Bergr to miss the duel while his challengers brave the extreme weather. Perhaps the terminology used reveals the narrator's overall disapproval towards the use of magic. As a result of the contradictions between the motive and following repercussions in both episodes, it is safe to say that the weather magic used in the passages from *Vatnsdœla saga* and *Finnboga saga* cannot be definitively classified as harmful or beneficial weather magic. The magical act can fall into either category, depending on whose perspective is considered.

The use of weather magic as a means to avoid death or injury is a common literary theme when it comes to efficatory weather magic, and in some scenarios, it can appear to be an act of beneficial magic. In the previously discussed episode from *Fóstbrœðra saga* (Ch. 10), Gríma employs weather magic to prevent her trustworthy slave, an innocent man, from being put to death. In this scenario, Gríma's motive for calling the magical wind is to allow her slave to make his getaway and avoid the punishment of death, after the slave killed a man to save Gríma's daughter. Gríma's motivation is to save the life of a man who protected her family. From her perspective, this can be seen as justification for her use of magic. On the other hand, her use of magic prevents a legal sentencing from being carried out and that could be interpreted as her evading the accepted punishment that holds her slave accountable for his violent actions. The narrative tone that is used alongside her weather-magic performance does not portray her magical skills as sinister in nature. As noted in the previous chapter, Bernadine McCreesh (2018: 72) suggests that the author of *Fóstbrœðra saga* consciously tries to distance Gríma's actions from a non-Christian practice that holds a negative connotation by using the phrase *fornu kvæði* (ancient poem) over a specifically magical term, like *galdr*. This careful choice in words can be seen as a way to portray a magical act as beneficial. However, it does not take into consideration every potential perspective on her use of magic. Gríma's magical act cannot be simply interpreted as either malevolent or helpful. While it is justifiably from her perspective, her use of weather magic goes against a punishment for a crime that would have been accepted in her society.

Although on the surface it does seem that beneficial weather magic appears in Old Norse literature, it is important to emphasise that the majority of surviving weather magic acts are portrayed in Old Norse saga literature, and these redactions carry a heavy Christian bias. Almost every example of weather magic found in these sagas is portrayed in a partly or wholly negative light, in terms of how the narrative context is presented. Magic in general reflects an extremely complex cultural phenomenon within the Old Norse corpus. Stephen Mitchell (2011: 77) suggests that magical tools like charms, incantations, and sacrifices are heavily linked to the worship of the non-Christian gods. Therefore, weather magic is naturally portrayed as a harmful act in post-Conversion literature because the performance of such magic in non-Christian settings must draw on the non-Christian gods.

The main scenario where contemporary medieval societies seem to accept weather magic as a beneficial tool is in the case of calming storms. As highlighted in the beginning of this section on efficatory magic, Bárðr's use of magic to counteract the magical storm that Helga summons in *Vatnsdœla saga* (Ch. 34) offers an example where weather magic stems from benevolent intent and the narrative context surrounding his actions mirrors those good intentions. Although published much later, in 1549, *Rerum Moscoviticarum Commentarii* (Notes on Muscovite Affairs) also presents a scene where calming a storm in Scandinavia is portrayed as beneficial (von Herberstein 1851–1852, 2: 106). This account, written in Latin by Baron Sigismund von Herberstein on the geography and history of Muscovy (a sixteenth-century Russian state), describes a local sailor performing an offering ritual to an unspecified power that successfully calms a storm at sea. After surviving a whirlpool off the coast of Vardø, Gregory Istoma and his men are stuck at sea for days due to high winds. The local sailor's ritual performance results in the storm abating and Istoma's ship finally being able to reach the fort. In this representation, the sailor's attempt to calm the extreme winds is portrayed as undeniably helpful.

There is a longstanding European tradition where magic practitioners who pacify severe weather are considered by general populations as being 'beneficial' or helpful towards the wider community. For example, as mentioned in the Introduction, Agobard of Lyon's *De Grandine et Tonitruis* (On Hail and Thunder, c. 815), a Latin treatise arguing against weather magic, describes villagers paying local *tempestarii* (weather-makers) to ensure their yearly crops would survive damaging hailstorms. Church authorities, however, rarely approved of individuals claiming to manipulate the weather unless the weather change was a result of praying and then classified as a miracle.[7] As early as 700, the Catholic Church persecuted individuals throughout Europe who were accused of manipulating the weather (Robbins 2015). Agobard strongly asserts that these *tempestarii* were false practitioners because surely the only ones with the ability to raise or calm the winds would have to be someone who has faith in God. Agobard explains that these individuals themselves would have not been able to influence the weather alone; they would only be able to achieve a climatic change by praying to God, who would in turn alter the weather on the believer's behalf. Authorities in the medieval Church were hesitant to accept anyone other than God, or perhaps a saint appealing to God, having such power over the natural world. In 1088, Pope Gregory VII wrote a letter to King Haraldr III of Denmark to demand that the king should no longer ascribe foul weather to his priests:

> There is one matter which has come to our knowledge and which cannot be passed over but must be sternly repressed by the apostolic prohibition, namely, that you ascribe to your priests the inclemency of weather, foulness of the air and certain ills of the body ... We therefore direct you by apostolic authority to abolish this pestilent practice absolutely from your kingdom and no longer presume to inflict such disgrace upon priests and clerics.
>
> *(Emerton 1967: 153)*

Here, Pope Gregory makes it clear that the Catholic Church would not acknowledge nor accept that even priests could have the ability to manipulate their natural surroundings.

There is also a fairly well-documented tradition of the general populace turning on these weather-magic performers once the intended magic fails to produce successful results or causes injury. In the case of the *tempestarii* described by Agobard of Lyon, the people of the Lyon region would violently turn against the weather-workers if their crops were still destroyed by a storm after the *tempestarii*'s magical performances. While direct evidence from medieval Scandinavia is lacking, the broader pattern of communities reacting with hostility toward failed weather magic suggests that similar attitudes may have existed in the North. This possibility is reinforced by Pope Gregory VII's letter, which reflects a concern over accusations of weather control. A later example from Early Modern Finnmark illustrates how these tensions could escalate, as a weather-magic performer faced retribution from his local community after a spell to bring favorable sailing winds went terribly wrong. In May 1627, a Saami man named Quiwe Baarsen was accused and tried for witchcraft. In the court documentation, the alleged weather magic is described as follows: On the Saturday before All-Saints day, the wife of the sailor Oluf Øresen approached Quiwe Baarson and promised him a payment of beer if he would raise the wind to ensure that her husband and the rest of the sailing crew could safely return home from sea. Quiwe agreed and performed the magical act, which gave the unintentional result of the winds becoming too strong and caused the boat to sink at sea.[8] The trial records note that Quiwe confessed that prior to this particular incident with the wife of Oluf Øresen, he often successfully summoned favourable winds for people. On 11 May 1627, Quiwe Baarson was found guilty of drowning five men via magical means and was sentenced to be executed by burning. Quiwe Baarson's trial demonstrates how a local community could consider a weather-magic practitioner beneficial for the community up until the point where the magical acts were no longer serving the intended purposes. As a result, the performers would be violently punished for their magical interference. Regardless of whether a certain magical act is seen as beneficial at some stage, it is inevitable that the same magical act can be interpreted as harmful at a later point. Even an act as seemingly straightforward as calming a storm or calling favourable winds is too multifaceted to place in a simple good versus evil taxonomy.

The cases from seventeenth-century Finnmark serve as an extreme yet historically connected continuation of anxieties surrounding weather magic that had been present in Europe and Scandinavia since the medieval period. The eleventh-century papal condemnation of weather-magic accusations, along with medieval Scandinavian laws regulating magical practices, laid the groundwork for the later criminalisation of weather magic. By the early modern period, these fears had intensified, particularly in regions like Finnmark, where minority groups such as the Saami, who were long associated with weather magic, became frequent targets of persecution. The Finnmark cases exemplify the culmination of earlier European and Scandinavian attitudes, demonstrating how concerns over magical control of the weather persisted and evolved into state-sanctioned violence against accused practitioners.

Conclusions

Many modern scholars categorise Old Norse magic within strict dichotomies, often framing it as either beneficial or harmful. However, this approach oversimplifies the nuanced nature of magical practices in medieval Scandinavian literature and often fails to acknowledge magic's multifaceted purposes. For example, while weather magic is sometimes depicted as a protective act, such as preventing injury or ensuring a safe voyage, the surrounding narrative tone does not always align with the performer's benevolent intentions. This suggests that societal perceptions of magic were not solely determined by the outcome of the act but were shaped by broader cultural attitudes toward magical intervention. The use of weather-magic-specific terminology, such as *gerningaveðr*, which predominantly appears in negative contexts, further underscores this disapproval, even when the magic was employed with seemingly positive intent.

Despite this prevailing ambiguity, certain instances of weather magic in Old Norse literature are portrayed as undeniably beneficial. Most notably, Bárðr counteracts Helga's magical storm, aiming to alleviate its severity. The overwhelmingly positive tone surrounding Bárðr's actions suggests that some magical interventions could be framed as virtuous, though these cases remain exceptions rather than the norm.

In conclusion, the depiction of weather magic in Old Norse literature reveals a complex interplay between intention, societal perception, and moral framing. The ambiguity surrounding these magical acts defies straightforward classification, highlighting the limitations of imposing rigid modern dichotomies onto medieval Scandinavian beliefs. Instead, a more contextualised approach that considers historical, legal, and narrative framing is necessary to understand the cultural significance of weather magic in the period. Furthermore, uncritically applying binary frameworks risks reinforcing the same biases present in medieval sources that were often used to marginalise certain groups, particularly non-Christian or Indigenous populations. For example, the demonisation of weather magic in early modern Scandinavia disproportionately targeted the Saami, whose perceived knowledge of wind and weather magic was often portrayed as malevolent 'witchcraft'. By perpetuating these medieval categorisations without interrogating their origins, scholarship risks replicating rather than analysing the cultural mechanisms that transformed certain magical practices into justification for persecution. A more nuanced approach is necessary, not only to better understand medieval magic but also to critically engage with the legacies of historical bias in modern research.

Notes

1 The name Heiðr is repeatedly used in Old Norse literature for women possessing divination skills. Examples of *vǫlur* with this name are found in *Vǫluspá*, *Orvar-Odds saga*, *Hrólfs saga kraka*, and *Landnámabók*. Simek (1993: 135) notes that

the name has been linked to a possible epithet for 'good' female practitioners of magic, but also suggests that it could just be a literary name for women, similar to Þuríðr.

2 Þorgríma *galdrakinn* is a magic practitioner known for her skills in weather magic and is discussed in the previous chapter on terminology. Although it is not clear what causes the magical weather, Þorgríma's ability to manipulate the weather and her rivalry with Þorgunna over their shared interest in Kjartan possibly suggests that Þorgríma was the source of the blood-rain (Livesay 1988:189).

3 For examples of this type of weather-magic altercation, see *Óláfs saga Tryggvasonar* (Ch. 63, Ch. 78, Ch. 80), *Óláfs saga helga* (Ch. 9), and *Færeyinga saga* (Ch. 31).

4 The pejorative nature of terms like *seiðmaðr* and *seiðkona* is evident in this scene from *Laxdœla saga*, where Kotkell and his family are described as *mjǫk fjǫlkunnigr ok enir mestu seiðmenn* (very skilled in magic and the greatest *seiðr*-practitioners). Their magical abilities are not framed as neutral or admirable but as dangerous and destructive, culminating in their summoning of a violent storm that drowns Þórðr and his men. The saga ultimately treats them as social outcasts, reinforcing the negative connotations of *seiðmenn*. Similarly, in *Gísla saga Súrssonar*, Auðbjǫrg, a *seiðkona*, uses magic to summon an avalanche as an act of vengeance, leading to the deaths of Bergr and his household. Her fate of being captured and stoned to death demonstrates that the saga scribe views her use of magic as both transgressive and punishable. These examples illustrate how Old Norse literature does not portray *seiðmaðr* and *seiðkona* as neutral or even ambivalent terms but rather as markers of socially and morally suspect individuals.

5 For example, *Fagrskinna* describes a famine that occurs under the reign of Gunnhildr's sons, where grains spoiled and fishing suffered (ÍF 29: 98).

6 'Svá mikil hríð kom á, at þat var með miklum ólikendum, hverr stormr var eða snájfall ... Veðrit helzt þrjár nætr, ok er upp létti, þóttust menn engan stað sjá snjóvarins' (ÍF 4: 311) ['A great blizzard came, such that a storm of snowfall like it would be exceedingly improbable ... the weather continued unabated for three nights and when it stopped people thought that not a trace of snow was to be seen ...' (CSI 3: 254)].

7 A comparison of weather magic and weather miracles in Old Norse texts will be explored in the fifth chapter of this study.

8 For a detailed description of Quiwe Baarson's magical act and an overview of the trial, see Hagen 1999: 58–60.

3

PERFORMING WEATHER MAGIC

Ritual, Speech, and Symbol

The performance of magic has long been recognised as a central component of medieval Scandinavian magical practices, yet discussions of weather magic have often prioritised outcome over process. While scholars have examined who performs weather magic and to what ends, less attention has been paid to how such acts are carried out within the narrative space of medieval texts. This chapter addresses that gap by focusing on the performative dimensions of weather magic as they are represented in Old Norse literature.

Drawing on Old Norse saga prose, poetry, and related narrative material, the chapter examines how weather magic is enacted through speech, gesture, movement, and ritualised behaviour. Particular attention is paid to the role of the spoken word, including chants, spells, and verbal formulae, as well as to bodily actions that frame magical performance as a public and socially legible act. By analysing scenes in which weather magic is performed, this chapter demonstrates how efficacy is constructed through performance and communal recognition. In doing so, it situates weather magic within broader medieval Scandinavian understandings of action, authority, and power, and lays the groundwork for later discussions of Christian reinterpretation and cultural continuity.

Sympathetic Magic

Richard Perkins (2001: 11) highlights that one method of performing weather magic is through an act of mimetic magic, also known as sympathetic magic. This type of magic is when an action is performed by a magic practitioner to mimic the desired effect that they are trying to produce. An example of sympathetic weather magic is demonstrated in *Reykdœla saga* (Ch. 14), when Ísgerðr wraps a cloth around her servant's head to cover their eyes from

DOI: 10.4324/9781003746058-5

the light, which creates an overall darkness and intense winds: a magical act that allows Ísgerðr's daughter to escape an unwanted marriage. Here, Ísgerðr's act of creating darkness by covering the servant's eyes results in a wider darkness appearing. In other words, like actions produce like results.

Eldar Heide (2006) describes a type of sympathetic weather magic where a magical wind is created by the spinning or twisting of an object. In this type of magic, the movement of the wind is created by a similar movement of an object (e.g. creating a whirlwind through the twirling of a spinning top). A clear example of this form of sympathetic wind-magic is found in a seventeenth-century account from Finnmark (*Tingbok for Finnmark* 1987: 169) when a magic wind is mustered by spinning an object. Additionally, there is a later folkloric tradition in sailing that spinning wheels were thought to be capable of raising dangerous winds, so sailors would not bring them onto a ship (Wikman 1917: 25). Heide (2006: 166) demonstrates the connection between summoning winds and a spinning motion by referencing the weather-magic scene in *Laxdœla saga* (Ch. 99). In this episode, the magic used to create a storm is referred to as 'harðsnúin frœði', or 'hard-twisted knowledge'. The terminology used to represent magic suggests that the magical knowledge that was necessary to create the wind was considered linked to the act of spinning thread.

Perkins' (2001: 11) brief overview of weather magic solely focuses on wind magic, and his exemplification of sympathetic weather-magic performance is limited to a folkloric tradition of sailors calling wind through whistling. However, there are no examples of sailors whistling to call the wind in Old Norse saga literature, nor in eddic or skaldic poetry. Instead, the most numerous examples of sympathetic weather magic found in these Old Norse sources use physical movement to mimic the stirring of weather conditions. Terry Gunnell (2014) notes that in many of these magical cases, the magic practitioner twirls or waves a goatskin or a similar cloth to bring a change to the weather, and terms this type of performance the 'goatskin twirl'. In the twelfth chapter of *Njáls saga*, the magic practitioner called Svanr performs the goatskin twirl to bring a fog to throw his enemies off course:

> Svanr tók geitskinn eitt ok veifði um hǫfuð sér ok mælti:
> 'Verði þoka
> ok verði skrípi
> ok undr ǫllum þeim
> er eptir þér sækja'.
>
> (ÍF 12: 38)

> Svan took a goatskin and wrapped it around his head and said:
> "Let there be fog,
> And let there be monsters
> And weird sights to those
> Who pursue you".
>
> (CSI 3: 17)

The amount of detail surrounding Svanr's magical performance is noteworthy in that the text reveals who the performer is, aspects of the actions used during the performance, and the exact magical words Svanr says. Variants of the goatskin twirl appear twice in *Vatnsdœla saga* (Ch. 37 and Ch. 47), as well as in *Víglundar saga* (Ch. 12) and *Harðar saga* (Ch. 25). Each of these portrayals include a similar amount of detail within the text, where the performer, action, and outcome of the magical act are revealed to the audience.

Gunnell (2014) also examines a second form of sympathetic magic that individuals are portrayed as using to manipulate the physical landscape, an act which he coins as 'magical mooning'. In *Vatnsdœla saga* (Ch. 26), this type of magic is performed when a woman called Ljót throws her clothes over her head, sticks her head between her legs, and walks backwards with a magical look in her eyes. Her magical performance is done with the intention to overturn the landscape and drive her enemies mad. The two other examples of magical mooning are found in chapter S 180 of *Landnámabók* and the fourteenth-century *Gull-Þóris saga* (Ch. 17). In each of these medieval Icelandic sources, acts of magical mooning are depicted as manipulating the natural world, but not as a manipulation of the weather. However, it is noteworthy that an Italian source from the fifteenth century does connect this type of performance to weather manipulation. During one of his sermons delivered in 1425 (*Le prediche volgari*), Bernardino of Siena described a diabolical way to protect crops from storms. Bernardino warns his audience of a version of magical mooning, where one would ward off a storm that was heading for their crops by bearing their naked rear-end to the approaching tempest. Bernardino cautions that this method of weather manipulation does have its drawbacks and tells the story of a local woman from Genoa who tried to chase away a storm with magical mooning: '[she] raised up her skirts from behind and pointed her rear end toward the bad weather. Just at that moment, lightning struck and killed her, because she had faith in such foolishness' (Mormando 1999: 97). Franco Mormando (1999: 97) remarks that much of Bernardino's work on the occult draws verbatim from William of Auvergne's (d. 1249) *De legibus*. Therefore, much of the information presented by Bernardino is over two hundred years old rather than from recent observation and is perhaps closer in time to the tradition of the Norse magical mooning.

Gunnell (2014: 150) suggests that although it is difficult to draw any concrete conclusions about the goatskin twirl and magical mooning, it is likely that these concepts stem from pre-Christian traditions that were preserved through oral transmission. Gunnell's conclusion derives from the lack of explanation surrounding the magical events portrayed in Old Norse saga literature, which he suggests could imply the following two possibilities: firstly, the saga compilers had no need to explain these magical activities in detail because the listeners or readers were already familiar with the referenced activity through an existing oral or written tradition; and secondly, if the audience did understand these magical references without explanation, then the magical act must have had some basis in actual practice or belief at some point. Yet, in comparison to the depictions of other types of magical events in Old Norse saga literature, even those of non-sympathetic weather magic, performances of the goatskin twirl and magical mooning tend to be some of the most descriptive magical acts. If we revisit the goatskin-twirl episode in *Njáls saga* (Ch.12), the narrative clearly depicts exactly who is performing the magical act, the actions that are performed to bring the fog, where the magical act takes place (outside of the home), the magical words that Svanr says to call the fog, and finally, confirmation that the magical act is successful in bringing the fog.

As demonstrated in the first chapter of this study, the use of weather-magic terms, like *gerningaveðr*, almost always occurs in Old Norse saga literature without any additional description of the performance used to change the weather. When the texts provide further information about the weather-magic event alongside the use of a weather-magic term (for instance, when the weather-summoner is revealed) the narrative never includes the magical actions used to manipulate the weather. I would argue that rather than focusing on these two specific types of performance, Gunnell's argument can instead be applied towards Norse weather magic in a more general sense. Although it is impossible definitively to trace any example of Norse weather magic back to a concrete pre-Christian practice, the sheer number of references to weather magic found in the written sources must have originated from a longstanding oral and written tradition. Furthermore, the majority of weather-magic acts that are represented in Old Norse saga literature and poetry are typically portrayed without accompanying details of the act itself, which Gunnell suggests is likely evidence that these traditions to some extent stem from pre-Christian practice. While it is difficult to assert with certainty, I agree with Gunnell's perspective that the lack of explicit descriptions of weather-magic performances in Old Norse texts might indicate that these practices were commonly understood and rooted in non-Christian traditions, thereby not necessitating detailed explanations within the narrative.

Wind Knots

Another means of producing wind magic is the act of tying knots onto a string or cord. A connection between wind knots and Saami weather magic became an increasingly prevalent motif in the early modern period. However, as Rune Blix Hagen (2017: 287) points out, neither the act of tying and untying knots to create magic, nor the more specific wind-knot motif, is unique to the Saami or Norse, nor is the tradition limited to the Middle Ages or early modern period. Aeolus from Homer's *Odyssey*, for example, gives Ulysses a bag full of wind, where a knot could be untied to open the bag and produce favourable sailing weather.

Perhaps the most well-known account of Saami wind knots comes from Olaus Magnus's 1555 account, *Historia de Gentibus Septentrionalibus*. In his explanation of these wind knots, Olaus Magnus claims that when sailors were stuck in Northern Scandinavia due to adverse weather, they would approach the local Saami communities to buy magical ropes that had three knots tied in them. The text describes the purpose of each knot as follows: the first knot was to be untied to create a gentle breeze; when the second knot was loosened, a stronger wind would appear; and finally, when the third knot was undone, such a gale would blow that the sailors would 'have no eye to look out for rocks from the bow, nor footing either in the body of the ship to strike the sails nor at the stern to guide the helm' (HGS I: 173).

Olaus Magnus' description of these wind knots raises the question of whether this type of weather-magic performance stems from an earlier, medieval tradition. Ranulf Higden's *Polychronicon* (Higden 1865), an English world chronicle from the fourteenth century, is the earliest surviving reference to Norse wind knots, which is likely copied from an even earlier anonymous work, *Geographica Universalis* (Livingston 2004: 33). In his description of the world, Higden notes that the island inhabitants of Wyntland have the reputation of selling wind to sailors in the form of knots tied onto a cord (Higden 1865: 320–329). Although Higden geographically pinpoints this barren island directly south of Denmark, his contemporaries and later scholars[2] debate the island's exact geographic location. Regardless, it is generally agreed that the island was considered to be located in proximity to modern-day Scandinavia. Wyntland's inhabitants' reputation for selling wind knots continued to spread in fourteenth-century England, as the motif is also included in two later fourteenth-century translations of *Polychronicon*,[3] as well as the anonymous fourteenth-century text, *Eulogium historiarum*. However, all of the accounts that include episodes of medieval Norse wind knots exist in sources that were written outside Scandinavia and there are no episodes of wind knots found in medieval Old Norse textual sources. The lack of Norse examples suggests that this wind-knot motif was not a common theme in weather magic originating in the Norse sphere; rather, the Northerner's reputation for being able to control the wind through the tying of knots was a motif constructed by an outside perspective. Tying the practice of wind knots to the Saami is equally problematic when relying on the sources mentioned above, and this will be discussed further in the following chapter on Indigenous Weather Magic. Perhaps there is a direct correlation between these fourteenth-century English works and Olaus Magnus' account of Saami wind-knot magic, and therefore the works may have contributed to the rising popularity of the early modern association between the Saami and wind magic, but the investigation of such a linear connection would fall outside of the scope of this chapter.

The Spoken Word

The linguistic evidence of the Old Norse weather-magic term *galdrahríð*, meaning magic storm, suggests that one way that weather magic, as well as *galdr* magic in general, was believed to be performed is vocally through songs or chants. As Raudvere (2002: 91) explains, when it comes to magic performance in medieval Scandinavia, the impact of the spoken word held a lot of importance, as it did in many other spheres of Norse life (e.g. in the poetic and legal spheres). Whatever was said during the magic performance could not be questioned, nor easily taken back, and it was almost as if the spoken words somehow transformed into something physical.[4] This emphasis on the spoken word can be seen throughout the magic tradition in the Norse world, including how weather magic is portrayed in the textual sources. Often the performance of weather magic is illustrated through a speech act, or something vocally expressed that not only presents information, but performs an action. The utterance of these spoken words become real when an authority performs them and their performance results in real consequences. In the case of weather magic, this authority figure is likely to be someone with a reputation for a knowledge of magic. An example of this occurs in *Fóstbrœðra saga* (Ch. 10), where Gríma, who is described using the term *fjǫlkunnigr* (someone with magic knowledge), recites *fornu kvæði* (old chants) to stop the winds from blowing.

Weather magic is often presented as a result of spoken words in Old Norse poetry. The final section of *Hávamál* includes a list of metrical spells, called *Ljóðatal*, or 'list of chants'. Carolyne Larrington (1993: 62) outlines this section as a list briefly describing the eighteen spells that Óðinn knows, but whose text is never provided. Larrington proposes that the *Ljóðatal* is an index of spells used to fulfil the literary functions of illustrating the different types of wisdom and unifying the different elements that make up the character of Óðinn. She further suggests that spells are not 'genuine' in that the real spells are never provided, in the same sense that Old English charms or the Old High German *Merseburg Charms* are. *Hávamál* is difficult to date with complete certainty, but McKinnell (2007) proposes that it could have been written down in the later tenth century.[5] A reference to weather magic is found in the 154th stanza, where the poem lists the eleventh spell that Óðinn knows, which allows him to quiet the winds and calm the sea:

154 Þat kann ec iþ
ellipta, ef mic
nauðr um stendr,
at biarga fari míno á floti:
vind ec kyrri vági á
oc svæfic allan sæ.
(Neckel and Kuhn 1962: 43)

I know a [eleventh] one if I am in
need, if I must save my ship
when afloat;
the wind I can quieten upon the wave
and lull all the sea to sleep.
(Larrington 2014: 34)

Although the actual words that are needed for Óðinn to manipulate the weather are not provided in the stanza, its inclusion in a list of chants or spells does suggest that there was a vocal element to the performance of this type of magical act.

McKinnell (2014: 134) explains that the first nine spells in *Hávamál*, including the weather-magic spell quoted above, stem from traditional charms that were already established in Old Norse poetry. McKinnell demonstrates this argument by noting that the nine traditional spells have parallels in *Svipdagsmál*, *Sigrdrífumál*, and *Rígsþula*, whereas the 'newer' spells that Óðinn is portrayed to know in *Hávamál* almost only have parallels in later sources. As McKinnell points out, *Hávamál* is not the only example of eddic poetry where a weather-magic spell appears in a list of *galdrar*. In stanza 11 of *Grógaldr*, one of the two poems that makes up *Svipdagsmál*, Gróa responds to her son's plea for help by giving him nine *galdrar* for protection against harm. The sixth spell that she lists is used to calm the seas:

11 Þann gelk
þér sétta, ef
á sæ kemr
meira an
menn viti, logn
ok lögr
gangi þér í lúðr saman,
ok léi þér æ friðdrjúgrar farar.
(Detter and Heinzel 1903: 185)

This sixth one I will chant for you: if you come to the
sea; running higher than men can stand,
may calm and quiet ocean run together for you in the mill-box
and may more peaceful voyages lie ahead of you.
(Larrington 2014: 258)

Grógaldr is a poem written in the *ljóðaháttr* metre and tells of a dialogue between Svipdagr, who has been given the task of finding and winning over the giantess Menglǫð, and his dead mother. The two poems that make up *Svipdagsmál* are not part of the Codex Regius collection,[6] and the

FIGURE 3.1 Olaus Magnus, On Women Skilled in Magic, from *Historia de Gentibus Septentrionalibus* (1555).

post-Reformation versions that still exist are often corrupt and appear to be removed from the archetype.[7] Although *Grógaldr* is a later addition to the corpus and we cannot confirm that the entirety of the story it tells existed before the present version, John McKinnell (2005: 203) suggests that there was at least a previous tradition of Gróa as a *vǫlva* and that she was known for her connection to magic. His argument is based on the use of the name Gróa, or variants of this name, alongside acts of magic in sources like *Skáldskaparmál* (17) and *Vatnsdœla saga* (Ch. 37). We also cannot securely say that Gróa was known for her ability to calm and quiet the seas prior to this version of the poem, but in *Vatnsdœla saga* (Ch. 37), Gró, a variant of the name Gróa, is also associated with a person who has the capability to exert a magical control over the surrounding landscape. In this episode, Gró employs sympathetic magic to cause a landslide that kills everyone in the house.[8] Furthermore, Gróa's weather-magic charm in *Svipdagsmál* having parallels in *Sigrdrífumál* and *Rígsþula*, alongside the examples of Óðinn calming the seas in *Reginsmál*, *Vǫlsunga saga*, and *Norna-Gests þáttr*, seem to suggest that it was likely that this type of weather magic was deeply rooted in a medieval Icelandic tradition.

Similar to the eleventh spell in *Hávamál* that is listed as a means for Óðinn to manipulate the weather, *Grógaldr* never provides the actual words to the weather-magic charm. As the earlier discussion of the term *galdrhríð* highlights, there are no surviving examples of *galdr* spells. Perhaps this is because the medieval scribes thought that these chants were too powerful or too dangerous to commit to manuscript. This emphasis on the power of the spoken word is, again, demonstrated in an episode from the legendary *Bósa saga ok Herrauðs* (Ch. 5), where Bósa's foster mother, Busla, confronts King Hringr of Götaland and threatens to curse him. The text continues to explain that, since the tale originated, it has become widely known that there are certain dangerous words that Christians have no need to know or say. Yet unlike the examples of weather magic found in *Hávamál* or *Svipdagsmál*, where the actual spells are not provided, *Bósa saga* disregards the inappropriateness of Busla's magical words and provides *Buslubœn*, or Busla's 'prayer'. Raudvere (2002: 134) proposes that the magic in Busla's prayer reflects common views about what a magic practitioner is capable of inflicting on their enemies, as many of the abilities listed in *Buslubœn* parallel the chants found in the other examples of eddic poetry(for example, *Hávamál* and *Grógaldr*). Yet the weather magic included in Busla's list of magical abilities is unlike the poetic examples of weather magic precisely because it is used as a means to inflict damage on an enemy. In her speech, Busla threatens to raise storms and turn all nature against Hringr. The weather magic described in the examples of eddic poetry above is used to quiet storms, which is preventative magic rather than offensive magic. The majority of the magic spells detailed in *Hávamál* and *Grógaldr* are reactionary in a way that focuses on aiding the performer, rather than actively inflicting something negative outwards onto an enemy.

Buslubœn is found in the manuscript AM 586 4° (c. 1450-1500), in *Bósa saga ok Herrauðs konungs*, a legendary saga that was most likely composed in the fourteenth century (SKALD 8: 25). Heizmann suggests that unlike the poetry that is normally included in Old Norse saga literature, *Buslubœn* '[does] not function to legitimise the narrative with metrically bound ancient lore but operate[s] as curse stanzas' (SKALD 8: 25), and the poem is constructed in a way that is similar to popular curse formulas (Heusler and Ranisch 1903). Heizmann (SKALD 8: 25) further suggests that although the title of the poem includes the Old Norse term *bœn*, meaning 'plea' or 'prayer', the noun should actually be considered synonymous with the Old Norse term *forbœnir*, meaning 'maledictions' or 'curses'. Although *Buslubœn* and the *galdrar* found in eddic poetry do share the inclusion of weather magic, the (possible) reference to *forbœnir* in the title of *Buslubœn*, alongside the harmful nature of the poem's contents, is more similar to the *Jarlsníð* in *Þórleifs þáttr jarlsskálds*.[9] The *galdrar* included in *Hávamál* and *Grógaldr* are separate to traditional curses in that they are not limited to being used as a means to inflict harm on others; rather, they are often employed as a means of protection for the individual conducting the spells.

Carving Runes and Using Amulets

The other means of performing weather magic that is described in eddic poetry is rune carving, which is demonstrated in the poem *Rígsþula*. This poem is not part of the Codex Regius, but is rather a part of the Codex Wormianus (AM 242 fol.), which is dated to the mid-fourteenth century. Yet based on the poem's metrical style and subject, it is normally included in the category of eddic poetry. In this poem, Rígr wanders the world and fathers the three social classes (slaves and servants, farmers and freemen, and nobles). The youngest son, a noble, inherits the title Ríg and becomes the first king. The title then is passed on to his son, Konr ungr. The poem offers a description of Konr ungr's knowledge of rune-lore, and the ability to quiet the seas is included among his magical skillset:

43 Enn Konr ungr kunni
rúnnar, Ævinrúnar oc
aldrrúnar;
meirr kunni hann mǫnnom biarga,
eggier deyfa, ægi lægia.
(Neckel and Kuhn 1962: 286)

But young Kon knew
runes, life-runes and
lifespan-runes;
and he knew how to help people,
to deaden sword-blades, quiet the ocean.
(Larrington 2014: 244)

This particular example of rune-lore, along with the chants from *Hávamál* and *Grógaldr*, emphasises the importance of controlling the weather while at sea. Each of the examples portrayed above describes a means to quieting the ocean, which would have been particularly useful for any culture strongly tied to sailing. McKinnell (2007) links Konr ungr's ability to calm the seas through rune-carving to the somewhat similar *brimrúnar*, however these runes do not have any direct manipulation over the weather. The *brimrúnar* described in *Sigrdrífumál* are instructed to be carved as a means of protection for ships sailing on tumultuous seas. In the poem's introductory narrative, Sigurðr wakes a Valkyrie from her deep sleep, who then presents Sigurðr with a list of helpful runes out of her gratitude. The fourth set of runes that the Valkyrie reveals is the *brimrúnar*, or 'wave-runes', which should be carved and burned on the rudder and oars for safe travel at sea:

10 Brimrúnar scaltu gera, ef þú vilt borgit hafa
á sundi seglmǫrom;
á stafni scal rísta oc á
stiórnar blaði oc leggia eld í
ár;
era svá brattr breki né svá blár unnir,
þó kømztu heill af hafi.
(Neckel and Kuhn 1962: 191)

Sea-runes you must cut if you want to have
guaranteed the sail-horses on the sea;
on the prow they must be cut and on the
rudder, and burnt into the oar with fire;
however steep the breakers or dark the waves,
yet you'll come safe from the sea.
(Larrington 2014: 164)

The exact runes that are needed to manipulate the weather or protect a ship at sea are not given in either *Rígsþula* or *Sigrdrífumál*, and perhaps this is for a similar reason as why the *galdr* spells are not committed to parchment: the magic that the runes represent was considered to be too powerful to write down. As for literary or material evidence of the *brimrúnar*, the longer version of *Flóamanna saga* contains a brief reference to a rune-inscribed oar, which Þorgils and his men find on the coast of Greenland.[10] There is also a runic inscription (N B249) found at Bryggen in Bergen which appears to be written on a miniature oar, which has been dated to c. 1332 (Liestøl et al. 1962: 106). However, Thomas Birkett (2016: 133) convincingly dismisses both of these examples as potential literary and runic analogues for the *brímrunar* by explaining that the Bryggen runes and the oar in *Flóamanna saga* appear to fulfil the function of relaying messages rather than offering a form of protection.

A few scholars (Perkins 2001; MacLeod and Mees 2006; Mitchell 2011) have explored the possibility that physical evidence of Norse weather magic exists, but the interpretation of these materials remains too circumspect to conclude with certainty that they are physical traces of medieval weather magic. Mitchell (2011: 66) briefly suggests that perhaps Scandinavian weather charms were a type of magic that left no archival or material footprint. This would be understandable given the emphasis on vocal performance in the portrayals of weather magic in Old Norse saga literature. Perhaps another reason for the lack of physical weather-magic evidence is because the only physical object described in weather-magic performances in Old Norse saga literature is the *seiðhjallr*. These temporary platforms are portrayed as being made of wood, which would deteriorate over time. Additionally, these structures are described as temporary and not elaborately decorated, which implies that the platforms would be unidentifiable even if we were to come across one in an excavation. A few archaeology scholars suggest that traces of *seiðhjallr* remains have been found in Uppland (Bäck et al. 2008; Arrhenius 2011), but as Olof Sundqvist (2012: 281) points out, this archaeological interpretation is problematic when compared to the literary sources, nor is it convincing from a broader religious phenomenological standpoint. The Old Norse textual sources normally describe the *seiðhjallr* as a temporary platform, mainly constructed with the purpose of resolving a temporary crisis or to host a travelling *vǫlva*'s divination ceremony. For example, the *seiðhjallr* described in *Laxdœla saga* (Ch. 35) is a temporarily constructed platform so that Kotkell and his family can perform weather magic through a *seiðr* performance to resolve a short-term issue (i.e. so that the family can seek revenge on Þórðr for accusing them of using magic). Sundqvist (2012: 2813) notes that descriptions from non-Norse sources also describe what might be examples of a *seiðhjallr*, but like the Old Norse sagas, none of these possible examples provide unequivocal support that a *seiðhjallr* was a permanent structure for worship or magic performance.

In their study of magic amulets and charms, MacLeod and Mees' (2006) interpretation and analysis suggests that two amulets bearing rune inscriptions are potentially surviving examples of weather magic. One such example is Sigtunablecket 2 (U AST1;166 M), a twelfth-century amulet found in Sigtuna. MacLeod and Mees (2006: 121) offer the normalised reading of the runes inscribed on the copper plate, which appear to call upon mist and (sun)shine to do the carver's bidding:[11]

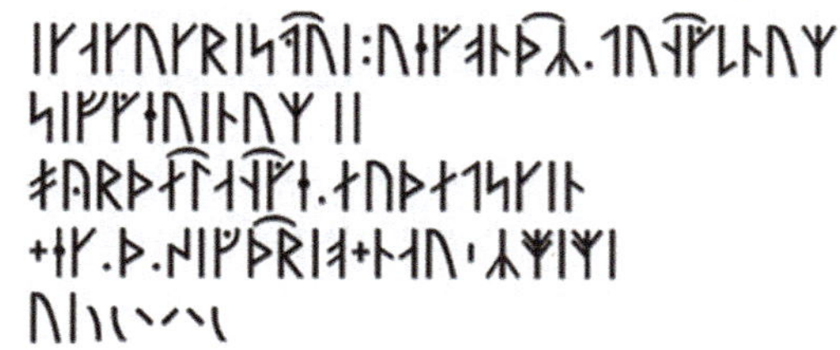

> Ik ak uk. Ris þū ī veg undiR tunglunum,
> sifgefnum! Ør þat angi! Eyð þat skīn! Ek þ(urs) seg þriu, nauðr nīu.
> Vīurr nān'k (?)
>
> '*Ik ak uk*. Rise and go away beneath the benevolent stars! Make this crazy mist! Destroy this, (sun)shine! I say three ogres, nine needs. (As) overseer of the sanctuary, I conjure (?)'.
>
> (transcription, transliteration, and translation from MacLeod and Mees 2006: 121)

Nordén's (1943:160) transliteration and translation, which MacLeod and Mees rely on for their analysis, are too disputed within the scholarly community to confirm that the amulet's runic inscription actually references weather magic. Firstly, the strong weathering on the amulet makes the inscription difficult to read with complete certainty. Furthermore, there is scholarly disagreement surrounding the direction in which the inscription is supposed to be read, as well as which language the runes represent. Nordén (1943: 160) dates the runes to the Middle Ages[12] and offers the first transliteration and translation of the runes in 1943, which is the interpretation that MacLeod and Mees use as a basis for their analysis. Also using Nordén's interpretation as a starting point, Gustavson (2014: 2–8) suggests that the inscription is written in Old Swedish rather than Old Norse. However, he diverts from Nordén's suggested reading order on side A of the amulet by rotating the artefact 180 degrees, while still maintaining the original reading order Norcén employed for side B, which Gustavson suggests was perhaps a possible inconsistency from the carver. Other scholars (Svärdström 1967; Källström 2010: 121) disagree with Gustavson's suggestion that the writing represents Old Swedish and argue instead that the rune inscriptions are reflective of Latin, whereas Pereswetoff-Morath (2017: 200) suggests that the inscription represents a combination of both Latin and Old Swedish.

Regardless of which reading direction is used for the transliteration or which language the runes represent, each of the proposed interpretations poses challenges in confidently linking the amulet to weather magic. MacLeod and Mees' translation of the phrase *'ør þat angi!'* as *'make that crazy mist!'* can be questioned, particularly with regard to the word *angi*, which in Old Norse typically refers to a smell or odour. However, the Old Swedish cognate *ange*, as attested in Söderwall's supplementary volume, can also mean *mist*, *fog*, or *haze*, and is attested in late medieval sources such as PMSkr 208.[13] This broader semantic range complicates the outright rejection of the 'mist' interpretation, though the term remains ambiguous and context-dependent.

Nevertheless, the link to weather magic remains tenuous. The translation of *skīn* to *(sun)shine* appears to rely on the assumption that *angi* (or *ange*) refers to atmospheric mist. While *skína* does mean *to shine*, and the phrase *sol skínn* would denote *sunshine*, the inscription lacks a clear reference to *sol* or any other explicit celestial term. Without such a reference, interpreting *skīn* as specifically related to sunlight rather than a general shining or glowing remains speculative. Furthermore, if the runes are meant to represent Latin rather than Old Norse or Old Swedish, the proposed connection to meteorological phenomena becomes even more uncertain. In short, while the semantic objection to *angi/ange* may not hold, the overall interpretation of the inscription as weather-related is still far from conclusive.

The second amulet that MacLeod and Mees (2006: 28) indirectly connect to weather magic is Öl SAS1989;43, an eleventh-century copper plate from South Kvinneby, located on the Swedish island of Öland. MacLeod and Mees provide the following interpretation of the runes carved on the amulet:

ᛯᛁᛯᚱᛁᛋᛏᛁᛦᚦᛁᚱᛒᛁᚱᚴ
ᛒᚢᚠᛁᛘᛅᛦᚠᚢᛚᛏᛁᚼᚢ
ᚦᛁᛋᚦᛅᛦᚢᛁᛋᛁᚾᛒᚱᛅᛚ
ᛏᛁᛚᚢᚠᚱᛅᚾᛒᚢᚠᛅᚦᚯᚱᚴᛅᛏᛁᚼ
ᛅᚾᛋᛘᛁᛦᚦᛅᛘᚼᛅᛘᚱᛁᛋᛅᛘᚼᚢᛦ
ᚼᛅᚠᛁᚴᛅᛘᚠᛚᚤᚠᚱᛅᚾᛁᛚᚢᛁᛏ
ᚠᛅᛦᛅᚴᛁᛅᚠᛒᚢᚠᛅᚴᚢᚦᛁᛦᚢ
ᚢᚾᛏᛁᛦᚼᛅᚾᚢᛘᛅᚢᚴᚤᚠᛁᛦᚼᛅᚾ
ᚢᛘ

Hǣr rīsti ek þǣʀ berg, Bōfi.
Mǣʀ fullty! Īhūð (?)
es þǣʀ vīss. En brā haldi
illu frān Bōfa.
Þōrr gǣti hans með þēm hamri sem uʀ hafi
kam. Flȳ frān illvētt! Fǣʀ ekki af Bōfa.
Guð eʀu undiʀ hānum auk yfiʀ hānum.

Here I carved for you (runes of) help,
Bofi. Help me! Knowledge (?) is certain
for you.
And may the lightning hold all evil away from Bofi.
May Thor protect him with that hammer which came from out of the
sea. Flee from evil! It (?) gets nothing from Bofi.
The gods are under him and over him.

(transcription, transliteration, and translation
from MacLeod and Mees 2006: 28)

Following the transcription, transliteration, and translation above, MacLeod and Mees suggest that, 'the carver of the amulet seems to appeal to the thunder god [Þórr] for security, presumably at sea, where sudden storms posed a very real danger without the protective deities hovering overhead and below' (MacLeod and Mees 2006: 29). Their analysis of this inscription, and the possibility that the amulet was made to provide protection from storms at sea, relies on previous academic work that connects the inscription to Þórr's mythological fishing expedition. Börje Westlund's (1989: 25–52) study was the first to propose a link between the amulet and the Þórr fishing myth. In the mythical tale, Þórr goes fishing with the giant Hymir and catches Jǫrmungandr, the Miðgarðsormr (World Serpent). In this scene, Þórr strikes Jǫrmungandr with his hammer, Mjǫllnir, before the serpent returns to the depths. This tale is one of the most substantiated pre-Christian Norse myths as it is represented in Old Norse sources such as *Gylfaginning* and *Hymiskviða*, appears in skaldic verse, and has been linked to the depictions on the Altuna stone, the Hørdum stone, the Gosforth stone, and Ardre VIII.[14]

Similar to the discussion surrounding U AST1;166 M, the transliteration and translation of the runes on Öl SAS1989;43 is also widely disputed in the scholarly community. Nilsson (1976) is the first scholar to offer an interpretation of the runes carved onto the South Kvinneby amulet, and it is in this original study where Nilsson first proposes that the amulet may have been used as protection for a sailor. Nilsson combines a reference to the sea,[15] the carving of a fish, which is also found on the amulet, and the location of where the amulet was found (Södra Kvinneby) as evidence that the amulet was used to invoke the gods for protection of Bove (the amulet wearer) while at sea.

Börje Westlund (1989) rejects some of Nilsson's interpretations and provides his own transcription and translation,[16] but maintains Nilsson's overall notion that the amulet was created as a form of protection while sailing. Westlund furthers this idea by suggesting that the inscription invokes the Þórr fishing myth from a Christian viewpoint, where invoking the pre-Christian gods can be used as a form of magic.

Following Nilsson's interpretation, Ivar Lindquist's transliteration and translations of the runes found on this amulet were published posthumously in 1987.[17] His analysis proposes that rather than a reflection of Þórr's fishing journey, the amulet's inscription is instead a poem dedicated to an earth goddess[18] and her son, Þórr. The poem references Ámr, who Lindquist suggests is a demon of sickness, and the inscription was meant to protect the wearer from illness. Lindquist's interpretation no longer associates the amulet with fulfilling the purpose of providing safe passage at sea, which would remove any possible links to weather magic. Although his reading has been criticised by Westlund (1989), Ottar Grønvik (1992) and Jonna Louis-Jensen (2001) later expand on Lindquist's interpretations of the amulet, suggesting once again that the inscription is a poem involving Ámr, who Lindquist originally proposed to be a demon of sickness.[19] McKinnell, Simek, and Düwel (2004: 66) mostly agree with Louis-Jensen's interpretation that the runic inscription would have been used as a curse against a demon that causes a skin infliction. However, their conclusions come with a slight variance on Louis-Jensen and Lindquist's reading of the name *Ámr* (the dark one) by likening the inscription to the giant *Ímr* (the sooty one).

MacLeod and Mees (2006: 29) build upon the foundations set by Nilsson (1976) and Westlund (1989) by suggesting that the amulet is perhaps similar to the magic associated with the *brimrúnar* from *Sigrdrífumál*, as well as a rune-inscribed staff that is used to calm dangerous waves, found in variant CCF 14 of the Faroese ballad, *Álvur kongur*.[20] Unlike the description of the *brimrúnar*, Westlund's interpretation of the runes carved on Öl SAS1989;43 does not specifically reference sailing or a ship, nor do the amulet's carvings align with the poem's instructions to either burn or carve the runes directly onto a ship.

The proposed connection between Öl SAS1989;43 and the rune-inscribed staff in *Álvur kongur* is equally problematic. The *Álvur kongur* variants were collected between the 1780s and 1880s, from a number of singers, and each of the variants can be rather distinct at points. The tale told in *Álvur kongur* is influenced by *Hálfs saga ok Hálfsrekka*, and the CFF 14 variant of the ballad bears the closest resemblance to the medieval saga's storyline (Mitchell 1985: 124). Yet, the ballad's episode of weather magic is only found in the CCF 14 variant and is not included in the medieval *Hálfs saga*. In the CCF 14 variant, the rune-inscribed staff is depicted as a tool that can be used to physically manipulate the weather by magic,[21] which saves a ship from capsizing during a storm at sea. Regardless of the amulet's connection with the Þórr fishing myth, its actual connection to weather magic, or its use as protection in the form of physically calming rough seas, is unconvincing. The only word that connects the inscription to the weather is the proposed reference to 'lightning', which is only found in Westlund's (1989) and MacLeod and Mees' (2006) interpretations. The inclusion of 'lightning' in the translation is not consistent among rune scholars, as Nilsson's (1976), Linquist's (1987), Grønvik's (1992), and Louis-Jensen's (2001) interpretations make no reference to lightning at all. McKinnell, Simek, and Düwel (2004: 66) suggest that Louis-Jensen's translation of 'the glistening one', should be read as a reference to Þórr's hammer, which they then associate with lightning. Even if one were to agree with Westlund's (1989) and MacLeod and Mees' (2006) transliterations and translations, the actual reference to lightning is not an indication that Þórr is manipulating the weather by conjuring lightning or sending it away; rather, the lightning itself (or Þórr's hammer, in the other interpretations) is invoked to protect the wearer. A translation including 'lightning' alongside the fish carving may point to an amulet worn for maritime protection, possibly invoking Þórr's power as a guardian figure at sea. However, this interpretation does not amount to clear evidence of weather magic in the form of direct, physical manipulation of natural forces, as seen in other literary or folkloric accounts.

Yet, it is precisely the ambiguity of such artefacts that is revealing. The absence of unambiguous weather-related terminology or narrative context makes the amulet's function difficult to define, but this lack of definitive evidence is itself significant. In this case, the inscription invites speculation about maritime protection and divine invocation but ultimately falls into the category of 'not proven' rather than disproven. As such, Öl SAS1989;43 does not provide firm evidence for weather magic, but it remains a compelling object that gestures toward the broader cultural logic of protection, divine power, and runic efficacy in medieval Scandinavian belief.

In his in-depth study on Þórr as a 'wind-raising' god, Richard Perkins (2001) suggests that the three figures possibly depicting Þórr, shown below, were used by Norse sailors as wind-amulets. He proposes that during the Viking Age, sailors would have used these figurines as a means to obtain a favourable wind for their sailing expeditions, or for unfavourable wind to thwart their competitor's sailing efforts (Perkins 2001: 75).

Figure 3.2 is made from walrus ivory, dating to the 'Late Viking Period' (Graham-Campbell 1980: 24–25),[22] and was found in Lund, Sweden. Ivar Lindquist (1963) links the figure to Þórr by combining the following three elements: Adam of Bremen's and *Fóstbrœðra saga*'s portrayal of Þórr as a seated figure, the effigy's distinct facial features, which Lindquist suggests is similar to those depicted on the Södermanland Runic Inscription 86's portrayal of Þórr (see Figure 3.3), and finally, the ring-and-dot pattern on the figure's back, which is similar to those often found on representations of Þórr's hammer.

FIGURE 3.2 Walrus ivory figure, Kulturen Museum (Lund) Image originally published in Kulturens Årsbok 1938: 43. Public domain.

FIGURE 3.3 Amber figure from Fedet (Roholte parish, Zealand). National Museum of Denmark. Photo: Lennart Larsen. Licensed under CC BY-SA.

FIGURE 3.4 The Eyrarland figure, Þjms. 892. National Museum of Iceland. Photograph by Kristján Magnússon. Used with permission.

The second effigy shown in Figure 3.4 is made from amber and was found at the Roholte parish in Fedet, Zealand, Denmark. Else Roesdahl (1992: 203) dates the figure to the tenth or eleventh century. Magnusson (1976: 68) and Perkins (2001: 68–70) link this carving to Þórr by likening its depiction to the god's well-known iconography (i.e., muscular, beard, large eyes, etc.). However, this iconography is very general, and the figure lacks unique features that would concretely tie it to Þórr, such as a hammer. The final effigy shown in Figure 3.5 is the Eyrarland Image, a figure cast of bronze that is also dated to the 'Late Viking Period' by Graham-Campbell (1980: 25). Although exactly who this figure is supposed to represent is disputed among scholars,[23] Richard Perkins (2001) connects the figure's features to those traditionally associated with Þórr. Yet once again, the iconography described is not unique to Þórr alone, and the figure lacks features that definitely tie it to the god.

Perkins (2001) hypothesizes that if these three miniaturised images represent Þórr, they existed among a tradition of effigies used to produce favourable or unfavourable winds for sailors at sea. He suggests that such images would have served as wind amulets, and perhaps to some extent, that Norse

FIGURE 3.5 Engraving of the Södermanland 86 runic inscription. From Stephens 1878: 33. Public domain.

sailors may have believed that the amulets' supernatural power may have been conferred on them by *blótskapr* (a sacrifice) or some other magical means. However, James Graham-Campbell (2003: 1371) notes that Perkins' conclusion does not acknowledge the existence of other strikingly similar medieval instances of beard clutching/pulling figures, both earlier and later in date and from entirely different European contexts.[24] These other figures do not represent Þórr, nor is it likely that they have anything to do with controlling the winds. Graham-Campbell suggests that the figurines above could fall into this category, which he concludes represent either virility or lust. Alternatively, Graham-Campbell (2003: 1371) proposes that the figurines above could be playing pieces, rather than cult items, and that the beard pulling would have then perhaps been a sign of pensiveness. Regardless, it is entirely possible that the three figures in Perkins' hypothesis do not stem from a single, unified purpose or motif. As is the case with the two runic amulets from Sigtuna and South Kvinneby, the link between the three 'Þórr' figurines above and a practice of calling favourable wind is too circumspect to confidently say that they were used for weather magic.

Offering Sacrifices for Favourable Winds

In medieval Scandinavia, the practice of offering sacrifices to divine beings to achieve a change in the weather was a common motif that is featured throughout Old Norse texts. Although the humans in these scenes are not directly producing the change in weather, relying instead on divine intervention, the concept aligns with Valerie Flint's definition of magic: '[magic is] the exercise of preternatural control over nature by human beings, with the assistance of forces more powerful than they' (Flint 1991: 3). In the case of human sacrifice, the magic is produced via the assistance of the gods or non-human beings (i.e. a force more powerful than the sacrificer). The following examination will illustrate how sacrifices in medieval Scandinavia, particularly to gods like Þórr and Óðinn, were considered a possible means to invoke favourable weather, underscoring the persistence of this belief during the medieval period.

In a scene from *Flóamanna saga* (ÍF 13: 274–282), Þorgils Þórðarson, a recent Christian convert, decides to immigrate to Greenland. When preparing to sail, the Old Norse god Þórr visits Þorgils in a dream, threatening to curse his trip with stormy weather. Þorgils ignores the dream and sets sail for Greenland, but the wind ceases as soon as the ship is out of sight from land. Þorgils discovers that there is an ox on board that has previously been dedicated to Þórr and has it thrown overboard, much to the protest of his crew. After a few rough months at sea, the ship and its crew wreck on the coast of Greenland. Richard Perkins (2001: 18) suggests that the manuscript's inclusion of the phrase *blóta Þór til byrjar*, or sacrificing to Þórr for a favourable wind (Perkins 2001: 18), likely represents genuine non-Christian terminology from pre-conversion Scandinavia. Perkins (2001: 18) argues that such an expression and the existence of other similar Old Norse phrases were remembered and used over the years due to their religious and ritual significance, reflecting real historical practices of sacrificing to the Old Norse deities like Þórr for favourable sailing conditions. While I am doubtful of the exact historical accuracy of this terminology and note that the use of these phrases as evidence of non-Christian practices is highly debated in scholarly contexts, its inclusion in medieval texts like *Flóamanna saga* indicates that the concept of sacrificing to Old Norse gods for favourable winds was in circulation during the time this saga was written.

Literary scenes featuring sacrifices to Óðinn in exchange for favourable weather illustrate the prevalence of the belief that sacrifices could influence natural forces. In Saxo Grammaticus's *Gesta Danorum*, the life of Starkaðr includes a significant episode where a raiding group, stranded by adverse sailing conditions, remains on land for almost a year. This situation is part of Óðinn's plan for Starkaðr to sacrifice his companion, King Víkarr of Norway. Starkaðr reveals to his men that a human sacrifice to Óðinn is necessary to secure favourable winds so that they can finally set sail. The crew agrees that the sacrifice should take place, casting lots that result in Víkarr's selection. Starkaðr deceitfully convinces Víkarr that the sacrifice will be symbolic, but ensures it is fatal. Following Víkarr's death, the winds change, allowing the fleet to sail. Although Saxo does not explicitly credit Óðinn with the weather change, the sailors' belief in Óðinn's power to grant favourable winds through sacrifice is evident, reinforcing the connection between sacrifices to Old Norse gods and obtaining suitable sailing conditions. This episode also has a close parallel in *Gautreks saga* (ch. 7), which similarly presents Víkarr as a sacrificial victim chosen to appease the weather. A comparative analysis of the *Gautreks saga* version and its significance is provided in a later chapter.

Additional evidence supporting the idea that weather could be influenced through sacrifices can be found in Dudo of St Quentin's *De moribus et actis primorum Normanniæ ducum*. As presented in the introduction section of the present study, Dudo provides a detailed account of a human sacrifice made to Þórr with the hope of securing good weather. In Dudo's account, he describes an event where a group of Norse raiders sacrifice members of their crew to Þórr, so that the god would change the weather and they would be able to sail to a new location. In the scene, Dudo explains that after a divinatory specialist determined that a human sacrifice must be made, the sacrificial victims, selected by lot, were struck on the head with a single blow from an ox-yoke:

> Cæterum, in expletione suarum expulsionum atque exituum, sacrificabant olim venerantes Thur, Deum suum. Cui non aliquod pecudum, neque pecorum, nec Liberi Patris, nec Cereris litantes donum, sed sanguinem mactabant humanum, holocaustorum omnium putantes pretiosissimum; eo quod, sacerdote sortilego prædestinante, jugo boum una vice diriter icebantur in capite; collisoque unicuique singulari ictu sorte electo cerebro, sternebatur in tellure, perquirebaturque levorsum fibra cordis, scilicet vena. Cujus exhausto sanguine, ex more suo, sua suorumque capita linientes, librabant celeriter navium carbasa ventis, illosque (v.l. deosque) tali negotio putantes placare, velociter navium insurgebant remis.
>
> *(Dudo: 129–130)*

> Moreover, as the last act on expulsion and departure (?), they used formerly to make sacrifices in veneration of their god Thor. To him they offered neither gifts of sheep nor cattle (?), nor of Father Liber (i.e. 'wine'), nor of Ceres (i.e. 'corn' or 'bread'), but they used to sacrifice the blood of human victims, thinking that to be the most valuable of all offerings. The procedure was this: When a divinatory priest had decided how things should be, the victims were struck viciously on the head with a single blow from an ox-yoke. And then, when the brain of each and every victim, selected by lot, had been crushed by a separate blow, he was spread out on the ground and the fibre (fibra) of the heart, that is to say the vein (vena), was carefully examined on the left-hand side (levorsum). After that, and in accordance with their custom, they would smear their own and their comrades' heads with the drained blood and promptly spread the sails of their

> ships to the winds, thinking to placate them (i.e. 'the winds'; or, accepting the reading deosque, 'the gods') by such actions. Then they would rapidly ply the oars of their ships.
>
> *(Perkins 2001: 20)*

Again, it is important to remember that Dudo was not an eyewitness to the events that he describes above, but his writing underscores the notion that in Dudo's time it was believed that Scandinavians used sacrifices and in this case, human sacrifices, to influence the weather. Whether or not this specific ritual occurred as described, Dudo's inclusion of the scene reinforces the idea that such practices were part of the cultural consciousness during his lifetime, even outside of Scandinavia itself.

One of the most infamous scenes involving sacrificial offerings to influence the weather comes from the literary depictions of the Battle of Hjörungavágr, traditionally thought to have taken place in 986 on the Sunnmøre coast of Norway. In this battle, Jarl Hákon, remembered as a devout believer in the Old Norse gods, led his forces against the Jómsvikings. Various sources recount the battle, with many attributing the decisive storm that turned the tide in Jarl Hákon's favour to divine intervention as a result from the sacrifice of Jarl Hákon's own sons.

In *Gesta Danorum*, Saxo Grammaticus portrays Jarl Hákon's desperate plea to the divine during the sea-battle, where he decides to sacrifice his own sons to invoke the help of unspecified gods:

> Quorum Haquinus perspectis copiis, cum intolerabile rebus suis onus imminere cognosceret, excipiendi eius materia non suppetente tanquam humane opis diffidentia diuinam amplexus, superos inusitato piaculo propitiandos curauit. Duos siquidem prestantissime indolis filios hostiarum more aris* admotos potiende uictorie causa nefaria litatione mactauit... [Dani] infestam nymbi uim, qua tanquam ab hoste lederentur, exciperent.
>
> *(Saxo 10.4.2: 690)*

> As soon as Hakon sighted these forces, he perceived that his situation was threatened by insupportable difficulties; as there was no provision to hand for encountering the danger, in his despair of human help he turned, supposedly, to divine aid, and set about appeasing the powers above with a rare propitiatory offering. Bringing his own two sons, both of outstanding ability, to the altar like victims, he slaughtered them in an abominable sacrifice to secure victory. ...[The Danes] met the hostile force of a thunderstorm, which injured them as sorely as if it were their enemy.
>
> *(Saxo 10.4.2: 690)*

This scene explicitly links Hákon's sacrifice to the change in weather, which in turn hinders the Jómsvikings. Saxo clearly presents Hákon's actions as deplorable, describing the sacrifice of his own children as an abomination.

Similarly, *Ólafs saga Tryggvasonar* from *Heimskringla*, which was composed between 1220–1230, portrays a scene where Jarl Hákon performs a sacrifice to an unknown divine being, which produces a magical storm.[25] The specific deity responsible for the change in weather is not mentioned in this source. In *Jómsvíkinga saga*, likely written in the late thirteenth century (JMSV: xvi), Jarl Hákon makes a sacrifice to the divine being, Þorgerðr, who then uses her skills in weather magic to change the battle's outcome.[26] Praying northwards, Jarl Hákon calls upon his patron goddess, Þorgerðr, offering various sacrifices until she agrees on the sacrifice of his son, Erlingr. Once the sacrifice is made, a magical storm appears:

> Ok síðan greiða þeir atróðrinn ok teksk þar nú af nýja inn grimmasti bardagi. Ok því næst tekr veðrit at þykkna í norðr ok dregr yfir skjótt. Líðr ok á daginn. Því næst flugu eldingar ok reiðar, ok því næst gørir á él mikit. Þeir Jómsvíkingar áttu at vega í gegn veðrinu. Þetta él var með svá miklum býsnum at menn máttu varla standask.
>
> *(JMSV: 35–37)*

> And thereupon clouds began to gather in the north and, as the day drew on, they soon covered the whole sky. This was followed by lightning and thunder, accompanied by a violent hailstorm. The Jomsvikings had to fight against the storm and the hailstorm was so fierce that men could hardly keep on their feet.
>
> *(JMSV: 36–38)*

As the battle resumes and the storm begins to settle, Jarl Hákon intensifies his plea to Þorgerðr, reminding her of the magnitude of his sacrifice. This results in an even fiercer storm, this time invoked by both Þorgerðr and her sister, Irpa.

Jarl Hákon's sacrifice is also depicted in the skaldic poem, *Jómsvikingdrápa*, composed around 1222 by Bishop Bjarni Kolbeinsson:

> 30 Hvervetna frák
> hǫlða, herr
> œxti gný darra
> gyr hreggviðum
> hjǫrva
> hrøkkva gunnar
> rǫkkum, áðr í ǫrva
> drífu
> ýtum grimmr at
> blóta, framm kom
> heipt en harða,
> Hǫkun syni tœki.

I have heard that everywhere men recoiled before trees of the storm of swords [(lit. 'storm-trees of swords') BATTLE > WARRIORS] bold in battle — the army caused the din of spears [BATTLE] to increase —, before Hákon, cruel to men in the snow-storm of arrows [BATTLE], proceeded at once to make a sacrifice; the harsh violence proceeded.

(SKALD 1: 984)

This source is particularly interesting in the context of the present study, because it specifically attributes Þorgerðr's ability to influence the weather to her knowledge of weather magic. In a following stanza in *Jómsvikingdrápa*, Hákon's sacrifice results in Þorgerðr, referred to in the poem as *Hǫlgabrúðr* (Hǫlgi's bride), exercising her skills in weather magic to stir a terrible blizzard with extreme hail, which sways the outcome of the battle:

32 Þá frák él illa
œða
Hǫlgabrúði,
glumði hagl á
hlífum harða
grimt ór norði,
þars í ormfrǫn
augu ýtum
skýja grjóti
(því knátti ben
blása) Barðid
hreggi keryðu.

Then I have heard Hǫlgi's bride [= Þorgerðr] stirred up the terrible blizzard; very cruel hail from the north resounded on shields, where the gravel of clouds [HAIL], driven by the storm, beat in the snake-flashing eyes of men; therefore wounds swelled.

(SKALD 1: 986)

Þorgerðr's weather magic is also portrayed in *Búadrápa*, a skaldic poem composed around 1200 by Þorkell Gíslason:

10 Ǫrum réð sér snǫrpum
— slikt vas raun
gǫrpum — flagð et
forljóta
af fingrum
skjóta.

Gerðisk grimt
fíkjum at
gumnum
ríkjum
— gnýr vas, hǭr
hlífa — hregg ok
loptdrífa.

The very ugly troll-woman shot sharp arrows from her fingers; such is a trial for brave men. An extremely savage storm and sky-blizzard arose against the mighty men; the clash of shields [BATTLE] was loud.

(SKALD 1: 950)

Although weather magic is used to sway the battle similarly to the scene in *Jómsvíkinga saga*, it is not depicted as an outcome of Jarl Hákon's sacrifice of his sons, which is not included in the text at all.[27]

These recurring depictions of sacrificial offerings to Old Norse gods in exchange for favourable winds and good sailing conditions underscore that the concept itself was prevalent in medieval Scandinavia, and the scribes were aware of the belief. The consistent portrayal of these rituals across various sources indicates a deeply ingrained cultural idea, reinforcing the connection between sacrificial offerings and the attainment of favourable weather during this period.

Galdrastafir, *Galdrabœkur*, and Old English Parallels

Arguably one of the most recognizable forms of magic from Iceland that is linked to bad weather is the well-known *galdrastafur* (magical stave), *vegvísir*, which cannot be directly linked to the Middle Ages. The compound name, *vegvísir*, comes from a combination of the Old Norse terms *vegr*, meaning 'a way' (Cleasby and Vigfússon 1874), and *vísa*, meaning 'to point' (Cleasby and Vigfússon 1874), and the first known example of this symbol is found in the Huld manuscript ÍB 383 4to, which is a collection of *galdrastafir* compiled by Geir Vigfússon around 1860. The symbol appears with the following instructions: 'beri maður stafi þessa á sér villist maður ekki í hríðum né vondu veðri þó ókunnugur sé (Handrit.is) [Carry this sign with you and you will not get lost in storms nor bad weather, even though in unfamiliar surroundings (own translation)]. The text surrounding the symbol indicates that the intended use of the charm, in combination with the second symbol below the *vegvísir*, is to help the user to find their way through rough weather. It is not a form of weather magic as defined in this study though, as its use does not directly manipulate the weather.

FIGURE 3.6 Vegvísir in the Huld manuscript, ÍB 383 4to (1860). Image from handrit.is. Manuscript preserved at the National and University Library of Iceland.

There are, however, two additional weather-magic staves intended to be used for the purpose of weather manipulation. These two staves belong in the manuscript Lbs 2413 8vo, which is dated to c. 1800. The first stave (no. 43 in Figure 3.7) includes the following instructions to conjure a favourable wind:

> Ef þú vilt fá byr. Rist þessa stafi á látún og gang í þá átt sem vindur skal úr koma og veif þeim þrem sinnum á loft þá sólin er í fullu austri svo kemur hagstæður vindur. Rist með stáli.
>
> (*Magnús Rafnsson 2008: 57*)

> If you want a favourable wind. Walk in the direction that you want the wind to come from and then carve these staves on brass and swing them three times in the air while the sun is directly in the east and a favorable wind will come. Carve with steel.
>
> (*Magnús Rafnsson 2008: 58*)

The second stave (no.126 in Figure 3.8) is used to call 'the northern weather'. The text included alongside the symbol instructs the performer to carve the letter on the *länguhöfuð* (head of a ling) and walk north while waving it in the air ('að gjöra norðaveður: Rist þennan staf á länguhöfuð og gang í norður átt og veifa í loft'[28] (Magnús Rafnsson 2008: 115)).

Although the *galdrastafur* tradition itself cannot be linked to the Middle Ages, two of the examples above do seem to reflect some aspects of weather magic that echo the medieval period. Stephen Flowers (1989: 11) suggests that the magic recorded in these *galdrabækur*, or Icelandic magic books, was developed from aspects of non-Christian magic, which perhaps appears to be the case with the weather-magic charms. Both of the staves from Lbs 2413 8vo inform the reader that the symbols themselves must be carved for the charm to be effective. As shown in the previous section on runes and amulets, the poem *Sigrdrífumál* also informs the audience that runes should be carved into the ship to ensure favourable winds. Although the symbols used in the *galdrastafir* are certainly not runes, the need to physically carve the charm to produce effective results remains. This could potentially highlight a continued or resumed belief that physically carving symbols was a method to perform Icelandic weather magic. Additionally, both of the sets of instructions above direct the magic-performer to wave the inscribed object around in the air, which is similar to the narrative examples described above that depict sympathetic weather magic. Once again, it appears that movements which mimic the wind can play a role in weather-magic performance, and this tradition possibly either carried on into, or was imitated in, the modern period. It is extremely difficult to demonstrate a continuity in these spells from the Middle Ages to the modern period, and it is entirely plausible that these

FIGURE 3.7 Stave no. 43 in Lbs 2413 8vo (c. 1800). Image from handrit.is. Manuscript preserved at the National and University Library of Iceland.

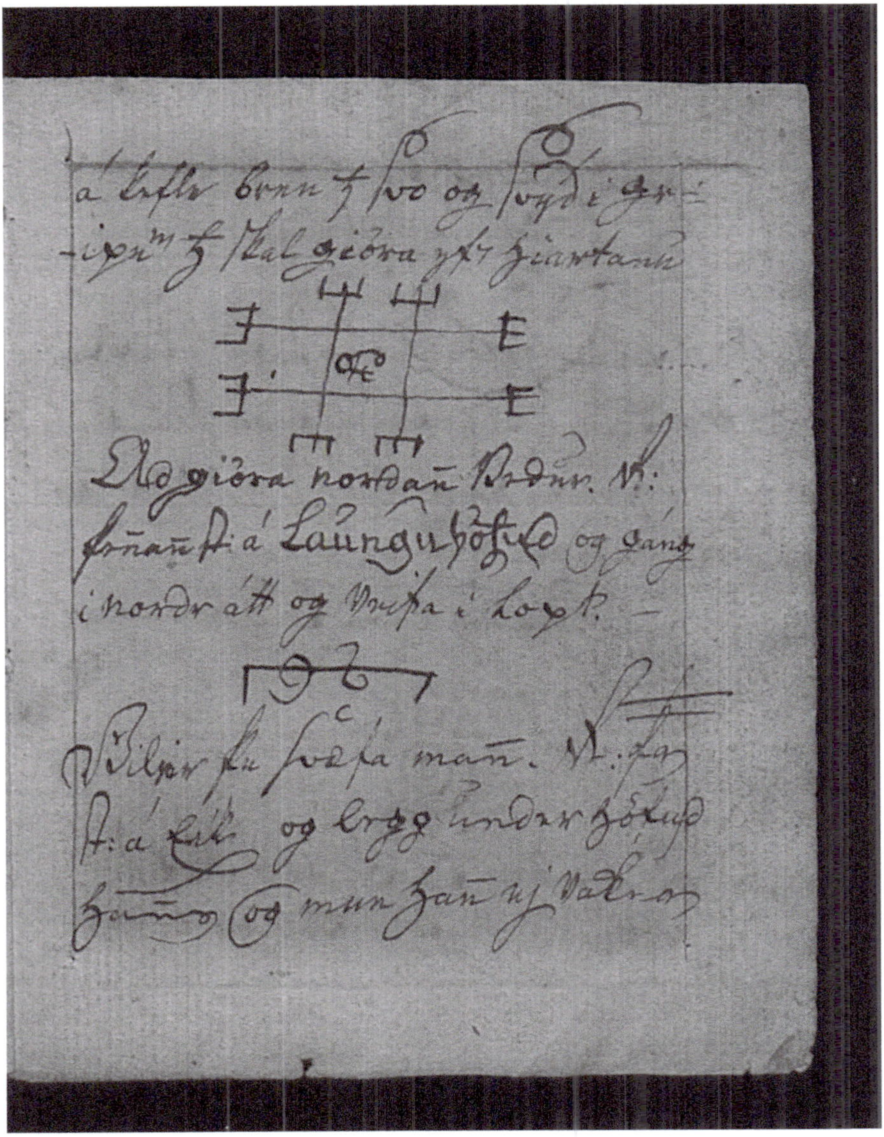

FIGURE 3.8 Stave no. 126 in Lbs 2413 8vo (c. 1800). Image from handrit.is. Manuscript preserved at the National and University Library of Iceland.

magic manuscripts were produced by nineteenth-century antiquarians who were interested in reconstructing Viking Age magic.

The nineteenth-century *galdrastafir* are preserved in compilations known as *galdrabækur*. References to these magical books in Scandinavia can be dated back to as early as the thirteenth century in Iceland, yet all of the surviving examples date closer to the early modern and modern periods. *Jakobs saga postula* gives one example of an early reference to a magic book, which dates from around the thirteenth century. In this tale derived from the *Legenda Aurea* of Jacobus de Voragine, Saint James the Greater converts the magician Hermogenes, who eventually offers his magical books and staff to the saint so that they can be burned. In the three of the surviving manuscripts that include this scene, two of the terms used specify that the books to be burned were magic: *galldrabocum* in AM 655–XII-XIII 4° (c. 1250–1275), and *fjǫlkynngisbók* in AM 630 4°ˣ (c. 1600–1700). The specific terminology used to specify magical books shows that, at the very least, the knowledge of magic books was circulating in medieval Iceland. Whether or not these books existed in medieval Scandinavia, or actually contained, or were thought to contain, charms or spells to prevent bad weather is another issue. It is important to note that the scene from *Jakobs saga postula* takes place in Judea, rather than in Scandinavia. This terminology used in this saga implies that medieval Icelanders were aware of magic books, but the episode is derived from a non-Norse source and takes place outside of Scandinavia. The use of *galldrabocum* in the thirteenth-century manuscript alone is not enough evidence to suggest that physical copies of these magic books were in medieval Iceland.

In *Jóns þáttr biskups Halldórssonar*, dated to the fourteenth century, there is an episode referencing a magic book, the reading of which eventually causes Jón to accidentally produce a powerful storm. This hagiographical tale follows the life of Jón Halldórsson, a Norwegian cleric who eventually serves as the bishop of Skálholt, Iceland, from 1322 to 1339. The scene in question takes place when Jón is a young man and has recently begun his studies in Paris. One day, his teacher leaves Jón alone in his study and Jón happens upon a magical text that has been left out. Once Jón reads from the text, a violent storm begins. The teacher returns to his study and comments on the ferocity of the storm outside, and it then occurs to him that the storm might have developed as a result of someone reading his magical book. Jón then confesses to reading the book while the teacher was away, and the teacher then opens the book and reads a section which eventually calms the storm. This particular scene reveals that in medieval Iceland, magical books from continental Europe were thought to possibly contain weather magic. Yet once again, the magical book in question is depicted outside of Iceland and therefore does not suggest the existence of such magical books in medieval Scandinavia. It is interesting that Jón Halldórsson comes across this magical practice while in Paris, where he is in the scholarly setting of his teacher's study. The Old Norse corpus often reflects this intimate connection between scholarship and magic, which was prevalent in medieval thought. An example of learned magic is highlighted in *Virgilessrímur* (thought to have been composed between c. 1300–1450), when the two main characters meet and discuss 'Bóka nám og brǫgðin ǫll' (Finnur Jónsson 1905–1922: 845) [book-learning and all kinds of tricks (Mitchell 2011: 83)]. Here, Stephen Mitchell (2011: 83) explains that by including additional terms associated with magic, the *rímur* clearly uses *bragð* as a reference to magic rather than just a mere sleight of hand. The connection between scholarly individuals dabbling in magic is not limited to a literary motif because as Mitchell (2011: 47) highlights, several late medieval Danish statutes prohibit clerics from practising magic because of its unquestionable ties to the devil.

Further evidence that magical books were circulating in the Scandinavian world comes from early fourteenth-century Denmark. Among the legal charges brought by King Erik Menved against Archbishop Jens Grand is the accusation that the archbishop possessed a *liber necromanticus*, or a book of necromancy (Mitchell 2015: 60). This suggests not only that magical texts were known in ecclesiastical and elite circles but also that their possession could be seen as politically and religiously dangerous. Although this charge emerges in a legal-political context, it supports the broader notion that magical books were not foreign to the Scandinavian imagination during the medieval period.

Medieval magic and scholarship are also tied together by the existence of medieval medical texts, or leechbooks, which mostly contained charms aimed at healing, protection and aiding everyday scenarios. Although there are no physical examples of magic books that can concretely tie the relationship between *galdrabækur* and weather magic to medieval Scandinavia, there are examples of Old English weather charms that are included in leechbooks. Two such weather-related charms are found in the *Old English Herbarium*, an Old-English medical text from c. 1000, which was translated from a fifth-century Latin work. This compilation lists important medicinal plants and describes various methods to use them in daily life. Although the work is not a magical text in the traditional sense,[29] it does include solutions, or charms, from outside the realm of healing. Among the many plants listed in the text are those that were favoured by seafarers, which were used to calm storms, avert hailstorms, or ward off dangerous lightning. The first entry of a weather-related charm is listed under aglaophotis. Among the flower's ability to cure fevers and cramps, or to ward off evil, the *Old English Herbarium* describes the following use for the plant: 'If anyone suffers a storm while rowing, take the same plant and set it on fire for incense; the storm will be held back' (Van Arsdall 2002: 223).

Although burning plants to change the weather does not appear in any of the Old Norse literary sources, its appearance in the Old English sources does echo the importance of maintaining some sort of control over the weather while at sea. The second plant in the text that is listed as a solution against harsh weather is *Ricinus communis*, or the castor-oil plant, which is used to control the weather both on land and at sea. For this entry, the *Old English Herbarium* outlines the multiple methods in which the plant should be prepared to ward off hailstorms:

> To turn away hail and storms, if you have the ricinus, or... [castor-oil plant] in your possession or you hang its seeds in your house or have it or its seeds in some place, it will turn back a hail storm. If you hang it or its seeds aboard a ship, it is surprising how it appeases all storms. You must pick this plant while saying this: *Herba ricinum, precor uti adsis meis incantationibus et auertas grandines, fulgora, et omnes tempestates, per nomen omnipotentis Dei qui te iussit nasci*. That is in our language: ricinum plant, I ask that you present at my song and that you turn away the hail and lightning flashes and every storm, through the name of almighty God, who caused you to be made. Also, you must clean when you pick the plant.
>
> *(Van Arsdall 2002: 225)*

Again, this example shows the universal importance of controlling the weather when living in a seafaring community. It is also interesting to note the vocal element that is required when one picks the castor-oil plants, which is similar to the importance of using a vocal element when performing Norse weather magic, as discussed previously. Even though this text is from outside Scandinavia, it comes from a region that the Scandinavians would have had much contact and cultural interaction with. Similar leechbooks focused on medicinal remedies also existed in medieval Scandinavia; however, I have yet to find an example of one that contains weather-magic charms. Given the Old English charms used to ward off storms, and the references to magic books in Old Norse literature, particularly that of the scene involving Jón and the magical storm, it is fair to say that medieval Scandinavians were at least aware that there was a tradition of weather-magic charms existing in either magic books or leechbooks outside Scandinavia. However, no evidence exists to suggest that Norse weather magic was associated with magic books or leechcraft. Instead, Old Norse textual sources reveal that Norse weather magic was more likely believed to be performed vocally, by mimicking gestures, or through the carving of runes.

Conclusions

The investigation into weather-magic performance within Old Norse society demonstrates the complexity of the beliefs and practices that are depicted in scenes of Old Norse literature, which illuminate the intricate relationship between medieval Scandinavians and the want and need to exert control over the natural world. Through the analysis of literary evidence of weather-magic performance, and the examination of archaeological materials that have previously been linked to weather manipulation, this chapter has contributed to the understanding of weather-magic performance as it is presented in Old Norse literature.

One of the foremost findings of this chapter is the prevalence of sympathetic weather magic in Old Norse saga literature. The narratives presented and analysed in the study above clearly depict individuals using symbolic actions or objects to exercise control over weather phenomena. For example, in *Reykdœla saga*, Ísgerðr wraps a cloth around her servant's head to create darkness and intense winds, allowing her daughter to escape an unwanted marriage. Similarly, in *Laxdœla saga*, the term *harðsnúin frœði*, or 'hard-twisted knowledge' (Heide 2006: 166), links the act of spinning to summoning winds, illustrating the concept of sympathetic magic. Terry Gunnell's (2014) notion of the 'goatskin twirl,' as seen in *Njáls saga*, where Svanr uses a goatskin to create fog, further demonstrates the role of symbolic actions in Old Norse weather magic.

Vocalisation also played a crucial role in Old Norse magical performances, especially in weather manipulation. The term *galdrahríð* ('magic storm') indicates that vocal performances, such as chants or spells, were integral to weather magic. In *Fóstbrœðra saga*, Gríma, described as *fjǫlkunnigur* (a knowledgeable magic practitioner), recites old chants to halt the winds. The poem *Hávamál* lists various spells known by Óðinn, including one that can quiet the winds and calm the sea. Similarly, in *Grógaldr*, Gróa chants to calm the seas for her son. These examples highlight that vocal elements were central to performing weather magic, transforming spoken words into physical reality.

Rune carving and the use of amulets were also portrayed in Old Norse texts as components of Old Norse weather-magic performance. The poem *Sigrdrífumál* instructs the carving of *brimrúnar* ('brim-runes') on a ship's rudder and oars to ensure safe passage. Archaeological evidence, such as the rune-inscribed amulet Sigtunablecket 2, which has previously been interpreted to call upon mist and sunshine, and the South Kvinneby amulet, which has previously been interpreted to invoke Þórr's protection against storms, would further support the importance of runes in magical practices (MacLeod and Mees 2006). However, the interpretation of these artefacts is contentious, with scholars debating the exact meanings and purposes of the inscriptions, and I do not agree that either amulet reflects weather-magic practices.

The practice of offering sacrifices to divine beings for favourable weather is also a recurring motif in medieval Scandinavian texts, reflecting the cultural belief in the power of such rituals. Dudo of St. Quentin describes a Norse raiding party sacrificing humans to appease Þórr for good sailing conditions, and Saxo Grammaticus's *Gesta Danorum* recounts Starkaðr sacrificing King Víkarr to Óðinn to change the winds. The Battle of Hjörungavágr features Jarl Hákon sacrificing his sons to invoke divine aid, resulting in a storm that turns the tide of battle, as depicted in *Gesta Danorum*, *Heimskringla*, and *Jómsvíkinga saga*. These narratives reveal the persistent belief that sacrifices to gods like Þórr and Óðinn could influence weather in medieval Scandinavia.

The existence of *galdrastafir* (magical staves) and *galdrabækur* (magic books) underscores the complexity of weather magic in Scandinavia. Although not directly tied to the Middle Ages, these symbols and texts reflect a continued belief in the efficacy of written charms and spells in Icelandic culture. The *vegvísir*, a well-known *galdrastafur*, promises to prevent the user from getting lost in storms, as described in the Huld manuscript from around 1860. The Lbs 2413 8vo manuscript, dating to around 1800, includes instructions for staves to conjure favourable winds or northern weather. These modern examples of weather magic perhaps echo older traditions of weather manipulation through physical symbols; a tradition which is represented in medieval Old Norse source material.

Old English sources provide valuable comparative perspectives on weather magic in terms of suggesting a potential *galdrabækur* tradition in medieval Scandinavia and their possible inclusion of weather-magic charms. The *Old English Herbarium*, a medical text from around 1000, includes charms to ward off storms using plants like aglaofotis and Ricinus communis. These texts highlight the universal concern with controlling weather, especially for seafaring communities. The vocal elements required in these charms, such as reciting incantations while picking plants, parallel the importance of spoken words in Old Norse weather magic. While there is no direct evidence of Old Norse weather-magic books, references in Old Norse literature, like *Jóns þáttr biskups Halldórssonar*, at minimum suggest awareness of such traditions.

Despite the multitude of rich literary descriptions of magic practitioners performing weather magic found in Old Norse literature, there is a notable absence of direct physical evidence for weather magic from the medieval period. Possible *seiðhjallr* remains in Uppland, as discussed by Olof Sundqvist (2012: 281), are debated and problematic in comparison to literary sources. The temporary nature of *seiðhjallr* platforms, used in rituals like those described in *Laxdœla saga*, means they would likely deteriorate over time, leaving little archaeological trace (Sundqvist 2012: 281). This absence suggests that many weather-magic practices were ephemeral or not intended to leave lasting physical evidence, aligning with the emphasis on vocal and performative elements in magic.

The examination of weather-magic performance in Old Norse sources provides a comprehensive understanding of the medieval Scandinavian's cultural understanding of how individuals could manipulate the weather. By analysing the specific examples from Old Norse texts and complementary continental texts, this chapter has offered insights into the beliefs and performances, as they are portrayed in literature, which shaped the Old Norse approach to weather manipulation. This exploration of interdisciplinary texts not only enriches our knowledge of Old Norse weather magic and performance but also highlights the broader cultural interactions that influenced these traditions. Through this lens, the chapter has demonstrated the intricate ways in which the medieval Scandinavians navigated their relationship with the natural world, seeking to harness the weather through performances of magic.

Notes

1 *Landnámabók* is preserved in three different medieval manuscripts that date to the late thirteenth century. However, scholars date the work's origin between the late tenth century to twelfth century, because the information within *Landnámabók* is used as a source for earlier works, like the twelfth-century collection of laws, *Grágás*. For more information on the dating of *Landárnabók*, see Sveinbjörn Rafnsson 2017.

2 For an overview of these different geographic proposals, see Livingston 2004. In this article, Livingston himself proposes that the island is actually a reference to Vínland. In the construction of his argument, Livingston (2004: 36) attempts to link Higden's inclusion of wind knots to anthropological surveys that report a connection between string figures and wind magic in North American Indigenous populations. However, these anthropological accounts are from the nineteenth century, and Livingston does not acknowledge the prevalence of Norse weather magic in sources that are contemporary to Higden's work.

3 John of Trevisa's translation in 1385 and the anonymous translation preserved in the fourteenth-century manuscript, MS Harley 2261 (Higden 1865).

4 For a general overview of the power of the spoken word in both Old Norse literature and ritual practice, along with a discussion of the differences between spontaneous speech or song and institutionalised forms of expression, see Raudvere 2005.

5 See also McKinnell 2007 for a survey of the scholarship and further discussions regarding the origins of *Hávamál*.

6 The Codex Regius, or GKS 2365 4°, is an Icelandic manuscript from the 1270s in which most of the Poetic Edda is preserved.

7 See Einar Ól. Sveinsson (1971–1973) for further discussion on *Grógaldr* and *Fjölsvinnsmál*.

8 'That afternoon, when the sun had set, a shepherd noticed that Groa went out and walked backwards round her house and said, "It will be difficult to resist the luck of the songs of Ingimund." She looked up at the mountain and waved a kerchief or cloth of hers in which she had wrapped much gold, and said, "Let whatever is fated come to pass." She then went inside and shut the door behind her. There was then a rock-fall on the house and everyone inside died' (CSI IV: 48).

9 See SKALD 8: 26 for Heizmann's analysis of the formulaic structure of *Buslubœn* compared to the *Jarlsníð* in *Þórleifs þáttr jarlsskálds*.

10 The oar carved with runes is found in the M-version of *Flóamanna saga*, which is primarily held in the manuscript AM 445 b, 4o (c. 1390–1425), along with some further fragmentary material in the manuscript AM 515, 4to (c. 1660–1695). Richard Perkins (1972: 433) suggests that the rune-inscribed oar was also present in the saga's original form.

11 Although the transliterations and translations presented in MacLeod and Mees' study (2006) heavily rely on previous scholarship, I have chosen to start the academic discussions surrounding both U AST1;166M and Öl SAS1989;43 by providing their transliterations and translations. This choice has been made because McLeod and Mees' scholarship approaches the two amulets with a clear focus on the magical properties and implications of the rune carvings. Additionally, the way in which they present their normalised readings is often more accessible to a wider audience, who may not be specialised in rune interpretations. The research proceeding and following the MacLeod and Mees interpretations will be considered in the text that follows each amulet's transliteration and translation.

12 Nordén's proposed date relies on his transliteration, where he identifies æ-runes in the inscription (Nordén 1943: 160).

13 I am grateful to Stephen Mitchell for drawing my attention to the broader semantic range of *ange* in Old Swedish, as glossed in Söderwall's supplementary volume, and for highlighting the relevance of this usage to the interpretation of the inscription.

14 See Sørensen (2002) for a survey of the Þórr fishing myth sources and its transmission.

15 Nilsson's translation of the inscription is as follows: 'Glory to thee bear I, Bove. Help me! Who is wiser than thee? And bear all in (the form of) evil from Bove. May Thor protect him with that hammer which came from the sea, (and which) fled from evil. Wit fares not from Bove. The gods are under and over him' (Nilsson 1976).

16 Westlund's translation of the inscription is as follows: 'Here may I carve (or I carved) protection for you, Bōfi, with… is sure for you. And may the lightning keep evil from Bōfi. May Thor protect him with that hammer… Flee from the evil being! It (?) will get nothing from Bōfi. Gods are below him and above him' (Westlund 1989).

17 Linquist (1987) offers two translations to the amulet's rune inscriptions. The first being, 'Here I, in poetry am familiar with the god(dess) beneath, for me, Bófi, to save myself. Earth, I am known to thee! May one son keep evil away from Bófi. May Þórr protect him with the hammer that smashes Ámr, the heavy Ámr. Flee, foul ill-wight! Get nothing from Bófi. Gods are under him and over him.' The second translation Lindquist provides is as follows: 'I here to Erka, the undergod of the world, for me, Bófi, to save myself. Earth, I am known to thee! And may the lightning raiser help evil from Bófi. May Þórr protect him with the hammer that smashes Ámr. Go the sea, Ámr! Flee, foul-ill wight! Get nothing from Bófi. Gods are under him and over him.'

18 The earth goddess is referred to as 'Erka', 'Fold', and 'Undirgoð' (the god below) in Lindquist's transliteration of the runes (Lindquist 1987).

19 Jonna Louis-Jensen (2005) offers a normalised text of the transliteration and the following English translation:

I cower herein,
under the god of soot;
I, Būfi, carry
a festering sore in my skin –
you know where the glistening one is –
keep evil from Būfi!
May Thor guard him
with the hammer
with which he strikes Āmr!
Be gone, evil being!
The affliction leaves Būfi,

there are gods below him
and above him.

20 *Álvur kongur* is the Faroese variant of a medieval Scandinavian ballad that is based on the same legendary story told in the Old Norse saga, *Hálfs saga ok Hálfsrekka*.
21 Weather magic is one of the more frequent magical motifs found in the Faroese ballad tradition. Examples of storms raised by magic can be found in *Artal kongur í Atlandi* (TSB E 161), *Álvur kongur* (TSB E 58), *Einars tættir* (TSB E 120), *Grimmars kvæði* (TSB E 98), *Grímur á Bretlandi* (TSB E 106), *Guttormur í Hattarmóti* (TSB E 39), *Jústinjalls kvæði* (TSB E 107), *Tiki-Álvs kvæði* (TSB E 22), *Sigmundar kvæði* (TSB E 30), and *Jómsvíkinga vísa* (TSB E 15). Although some Faroese ballads may stem from Old Norse saga literature, the previously listed ballads will not be discussed in depth during the present study due to their late collection date. For an overview of the weather-magic motifs found in each of the ballads above, see Korecká 2017.
22 Graham-Campbell (1980) defines the 'Late Viking Period' as from the second half of the tenth century to the beginning of the twelfth century.
23 See Perkins 2001 for a detailed summary of each theory.
24 Cf. the beard-pulling figures that decorate the Book of Kells manuscript, or the beard-pulling man who appears in the canon tables of the Barberini Gospels (Graham-Campbell 2003).
25 'It was rumoured that in this battle Jarl Hákon had made a sacrifice of his son Erlingr for victory, and after the storm blew up, and then the casualties turned against the Jómsvikings' (*Heimskringla* II: 178).
26 An almost identical scene is depicted in the version of *Óláfs saga Tryggvasonar* found within *Flateyjarbók*.
27 Although the poem implies weather magic, it is not explicitly clear whether Þorgerðr is the one responsible. John McKinnell suggests that the 'ugly troll-woman' could be either Þorgerðr or her sister Irpa, though the reference to ugliness might lean towards Irpa (McKinnell 2002: 284).
28 To call up a northern blizzard. Carve this stave on the head of a ling and walk northwards and wave it in the air.
29 Although Cockayne's original translation (1864) focuses on the magical aspects of this text, and therefore dismisses its medical relevance, Van Arsdall (2002) convincingly places the text into a unique medieval medical tradition, one which draws on elements from classical sources. For an overview of Cockayne's original interpretation and Van Arsdall's reassessment of its contemporary context, see Van Arsdall 2002.

PART II

Weather-Magic Performers

4

INDIGENOUS WEATHER MAGIC

Saami, Bjarmians, Kvens, and Norse Exchange

As Charlotte Damm (2005) notes, Indigenous groups all over the world have often been cast into the role of the 'powerless other', to uphold the Western colonial dichotomies that reinforce the dominance of those in power. In Old Norse texts, however, the Indigenous populations of Fennoscandia are often represented as having power to influence the course of the narrative, especially through the use of magic. For example, in the previously discussed weather-magic scene in *Ketils saga*, a coastal Saami community calls a storm that affects Ketill's sailing course, which eventually brings him to meet and defeat the Saami king Gúsir.

Regardless of their prominent roles in these medieval narratives, the Indigenous characters represented throughout these Norse texts are often treated by scholars as 'mysterious' or peripheral actors who support a Norse narrative, without having any actual influence on Norse society (Wang 2024). Eleanor Rosamund Barraclough (2017: 28) points out that an overwhelming amount of literary scholarship that focuses on the Saami tends to concentrate on what has previously been interpreted as cultural features that separate the Saami from the Norse, such as their peripheral location in the 'Far North', the continued practice of their pre-Christian religion, and their knowledge in magic. Only recently has the focus on Indigenous groups as active contributors to the historical and literary narratives taken more precedence in the wider study of Old Norse texts, and these assumed cultural separations are being dispelled.

Recently, there has been a call for scholars to challenge these problematic perceptions of a monocultural Fennoscandia and instead showcase the

DOI: 10.4324/9781003746058-7

cultural fluidity in medieval Fennoscandia (Wang 2023). As an example of how these problematic interpretations are being overturned, Solveig Marie Wang (2023) explores the socio-cultural fluidity between the Norse and Saami, regardless of political or spatial borders, while also demonstrating a medieval Saami presence in more southerly contexts. Furthermore, Else Mundal (2018: 18) addresses the issue of viewing Saami characters as the 'heathen neighbours' by reminding scholars that all Saami did not uniformly convert to Christianity in the seventeenth century. Rather, their conversion process began around the same time as the Norse conversion, and was a drawn-out development. This chapter aims to focus on the Indigenous magic of Northern Fennoscandia as a case study to showcase the influence that Indigenous communities had on Norse magic and to demonstrate that there was a cultural awareness of regionality in terms of the various weather-magic practices in medieval Fennoscandia.

One of the main challenges of studying magic in medieval Fennoscandia is that the surviving representation of this magic solely exists in sources written from a Norse perspective, which cannot fully represent the medieval Indigenous worldviews or actual practices. To approach these sources for a completely accurate representation of the medieval Saami viewpoint is, of course, an impossible undertaking due to the absence of first-hand accounts written by these communities. Furthermore, academic scholarship focusing on Indigenous religions has a tendency to use problematic terms to designate religious practices, ceremonies, and knowledge. As Konsta Kaikkonen puts it:

> An exoticist fascination towards the age-old beliefs and practices of the [Saami], that existed previous to as well as alongside Christianity, has often focused the interest of majorities in the Nordic states – and beyond – towards the [Saami]. This has resulted in several written descriptions of [Saami] religion... [having been] translated according to the prevailing zeitgeist as 'witchcraft', 'sorcery', or 'idolatry' before the 18th century; 'superstition' in the times of the Enlightenment; and 'nature religion', 'primitive religion' or 'shamanism' in the 19th and early 20th centuries.
>
> *(Kaikkonen 2020b: 2)*

This continued interest in Saami traditions and a long tradition of Western scholars defining and describing Indigenous practices has shaped, in part, how the Indigenous groups of Fennoscandia have been viewed and treated by non-Indigenous majorities (Kaikkonen 2020b: 2). As demonstrated in the previous chapter on magic terminology, I agree with Porsanger's (2018) and Kaikkonen's (2020a) call for scholars to avoid negatively charged terms, like 'witchcraft', 'sorcery', 'witches' and 'sorcerers', that are often used to exoticise both Norse and Indigenous ceremonies, acts, and practitioners.

Weather Magic or Religious Rituals?

Archaeological and textual evidence suggests that the Saami and Norse populations shared certain central religious concepts and practices (Mundal 1996; Solli 2002; Price 2002, 2019; Hansen & Olsen 2014). In a pre-Christian context, these shared religious foundations may have formed the basis of a mutual understanding and respect between the different cultural groups of Fennoscandia, whereas a divide between these religious practices, and therefore deterioration of shared understandings between cultural groups, most likely occurred following the Norse conversion to Christianity (Hansen & Olsen 2014: 50). Strömbäck (1935) suggests that Norse *seiðr* was directly influenced by Saami rituals. Kusmenko (2009: 5) further argues that even the term *seiðr* could be a loan word from Finno-Ugric languages, as the many Finno-Ugrian parallel terms, like the North Saami word *sieidi* (a site for ritual offerings), might suggest.

Weather manipulation and weather divination are portrayed as two of the shared features of both Norse and Indigenous groups of medieval Fennoscandia. *Hversu Noregr byggðist*, a text about the origins of various legendary Norwegian lineages that is preserved in the late fourteenth-century *Flateyjarbók* manuscript (GkS 1005 fol.), includes an example of a region attempting to ensure good weather through sacrificial offerings. In this source, Þórri, the son of Snær, is portrayed as the ruler over Kvenland and Finland. As ruler, Þórri receives sacrifices from the Kvens so that he would bless the region with good winter weather (HveNo 1954: 75). The offering scene from *Hversu Noregr byggðist* mirrors two ritual offering scenes from *Ynglinga saga*, in which sacrifices are made to influence seasons. In Chapters 9 and 10 of *Ynglinga saga*, Njǫrðr and Freyr both receive offerings from the Svíar in exchange for good harvests. It is said that the harvests were so good under Njǫrðr's reign that he was thought to have influence over the seasons. Later, in Chapter 15, the Svíar continue to perform sacrifices to influence harvest seasons. Under the reign of Dómaldi, there were multiple famines that led to the sacrifice of livestock and humans, which eventually escalates to the sacrifice of the king himself, as an attempt to improve the harvest output. These two Old Norse texts reveal that performing offering ceremonies was thought to be a shared practice that took place in multiple cultural groups in Fennoscandia.

When studying the many examples of Indigenous weather magic that are found in Old Norse texts, scholars often consider the question of whether these magical acts represent a real religious tradition. However, I would argue that it is impossible to prove that the textual representations of Indigenous magic, or even Norse magic, depict actual real-life religious practices. Indigenous weather magic is similar to Norse weather magic in that it would have not left any clear physical traces that indicate weather manipulation with any certainty.[1] While Price (2019: 217) suggests that counteracting environmental crises, such as difficult weather and storms, fell under the main ritual territory of the *noaidi*, or Saami ritual performer,[2] there is no direct evidence for this type of weather manipulation ritual outside early modern and modern sources. Many descriptions of weather magic as a Saami religious performance are found in court records detailing witch trials, which cannot be trusted to be completely accurate portrayals of these ceremonies or acts due to the traumatic conditions in which the confessions were provided. For example, in May 1627, the Saami *noaidi*, Quiwe Baarsen, was accused of using wind magic that resulted in the death of five fishermen.[3] During his trial, as described in the court's account, Baarsen told the court that he was often asked to control the winds and, similar to the 1692 trial of a Saami man named Poala-Ánde (Anders Poulsen), his outline of the use of drums includes the ability to predict the weather and to produce favourable winds for sailing.[4] Both of these men gave their descriptions of weather-magic performances during trials, which they knew could, and in the case of Quiwe Baarsen did,[5] end in a death sentence. In her recent article on the trial of Poala-Ánde, Liv Helene Willumsen (2020) argues that these early modern witch trial documents should not be taken at face value, as many scholars have previously done.[6] Willumsen demonstrates that the clear linguistic barriers of the trial and Poala-Ánde's tailored descriptions of the drum symbols and rituals to appear as a believer in the Christian god reveal that considering the traumatic context of these trials is crucial when interpreting the religious descriptions provided in these documents. In Poala-Ánde's case, and indeed for other accounts involving *noaidi* performances, the uncertainty of the trial outcome and desire to be found innocent influences how the accused describe religious ceremonies and acts (Willumsen 2020: 154). Regardless, even if either Baarsen's or Poala-Ánde's testimonies were an accurate depiction of Saami rituals and religious ceremonies, there is no surviving evidence from the Middle Ages or earlier that connects ritual performances to the *noaidi*, drum performances, and weather manipulation. Given the lack of direct evidence from the Middle Ages or earlier connecting Saami rituals, the *noaidi*, drum performances, and weather manipulation, we cannot definitively conclude that the depictions of Indigenous weather magic in Old Norse saga literature accurately reflect historical *noaidi* practices. Instead, the scholarly tendency to connect these literary depictions to *noaidi* practices might be influenced by later interpretations or misunderstandings of Saami traditions.

Although the 1627 Saami case is the only trial examined in detail here, it is not the only post-medieval example in which weather magic was treated as a serious and dangerous offence in Scandinavia. Particularly relevant are the Danish and Norwegian witch trials associated with the voyage of Anne of Denmark to Scotland in 1589, where several women were accused of raising storms to impede her crossing. These trials, which unfolded in Copenhagen and elsewhere, offer some of the most direct historical examples of weather-magic charges in early modern Scandinavia. While these cases fall outside the temporal focus of this study, their relevance to the cultural and legal perceptions of weather magic in Scandinavia cannot be ignored. For this reason, they are briefly revisited in the final part of the book.[7]

In the previously discussed scene from Sigismund von Herberstein's *Rerum Moscoviticarum Commentarii* (1549), the included act of weather magic is perhaps an example of an outsider's interpretation of an Indigenous religious practice. In this episode, a local sailor's ritual of offering oats and butter on a stone clears the stormy weather, which allows Gregory Istoma and his crew to continue their sailing route. While the ethnicity of the sailor that performs the offering is never established in the text, Perabo (2016: 122–123) suggests that because the source implies that the individual is local to the area east of Vardø, it is likely that they are of Saami background or belonged to another Indigenous group.

Although I am wary of using early modern sources as descriptors for medieval rituals or ceremonies, Lyonel Perabo (2016: 122) notes that there are interesting similarities between this scene and medieval depictions of individuals or communities providing religious offerings to change the weather. For example, he (Perabo 2016: 123) highlights its likeness to the previously mentioned offering episode found in *Hversu Noregr byggðist*. Although details surrounding how the sacrificial ceremony is performed are not revealed in *Hversu Noregr byggðist*, the overall outcome of weather manipulation resulting from a sacrificial act remains the same. Perabo (2016: 123) also proposes that the weather-magic episode in *Sturlaugs saga* mirrors the offering scene in *Rerum Moscoviticarum Commentarii*, although I am not convinced by his argument. In this early fourteenth-century episode, Sturlaugr and his crew help a woman murder her sisters in exchange for two days of good sailing weather, while they are seafaring off the shores of Finnmǫrk. However, unlike the previously discussed scenes in *Rerum Moscoviticarum Commentarii* and *Hversu Noregr byggðist*, the good sailing weather in *Sturlaugs saga* is the result of a dramatised bargain with a magic practitioner (*trǫllkona*) and seems less likely to be an example of a religious offering ceremony. In this episode, Áki, the son of Jarngerdr, rows out to sea alone, off the coast of Hálogaland, and meets a lone woman, who is described as a *trǫllkona*. She bargains with Áki by asking:

> flytir mik til eyjar þeirar, er hér er skammt frá landi. Þar hefir látizt faðir minn frá miklu fé, en vér erum þrjár systr ok eigum arfi at skipta með oss. Vilda ek fyrri koma en þær. Mun ek gefa þér tveggja dægra byr, þá er bezt kemr við.
>
> *(FSN 3: 132)*

> take me to that island, a short distance off the coast. My father has died there, leaving a lot of money. We are three sisters and we have to divide the inheritance between us. I want to come out ahead of them. I will give you a day and a night of favourable winds when it will be most useful.
> *(Waggoner 2014: 152)*

Later in the episode, Framarr, one of Áki's sailing companions, rows the same boat out and meets another woman named Hildr, who offers him the same exchange – a row to the island in exchange for favourable winds. This weather-magic episode in *Sturlaugs saga* does not mirror the religious offering scenes in *Rerum Moscoviticarum Commentarii* and *Hversu Noregr byggðist* because there are no descriptors in the text that indicates the *trǫllkona* made a religious offering to change the weather. Instead, the narrative uses weather magic as a skill that the *trǫllkona* can offer to bargain with Áki. Furthermore, the absence of descriptors indicating a religious offering made by the *trǫllkona* to change the weather distinguishes this episode of weather magic from the scenes in *Rerum Moscoviticarum Commentarii* and *Hversu Noregr byggðist*. Thus, the narrative of *Sturlaugs saga* portrays weather magic as a bargaining tool rather than a religious ritual, challenging the notion of it mirroring the two religious offering scenes.

Rerum Moscoviticarum Commentarii's inclusion of a sacred rock makes it plausible that the local sailor belonged to one of the Indigenous communities. Furthermore, the sailor uses the word 'Semes' (Von Herberstein 1851: 107) when describing the rock upon which he made the sacrifice, which I argue is perhaps an adapted spelling of the Saami word *Šeme*, derived from *šämä*. Hansen and Olsen (2014: 36) state that *šämä* is related to the Baltic-language words *žēme* or *zeme*, which shares the origin with the equivalent Russian word *zemlya*, meaning earth or land. In the Baltic region during the same time period that *Rerum Moscoviticarum Commentarrii* was written, *žēme* or *zeme* can specifically refer to sacralised earth or the goddess of the earth (Vaitkevičienė 2011: 107). I therefore suggest that it is plausible that the 'Semes' referred to in the text could refer to a sacred piece of earth, where the sailor's sacrifice is made. Although it is not specifically designated as such, the offering depicted in *Rerum Moscoviticarum Commentarii* appears to be reminiscent of a *sieidi*, which is a sacred place of sacrifice, often a rock of unusual shape, which is prevalent throughout the regional variations of Saami religion.[8] Archaeological material from the Iron Age and early Middle Ages shows that *sieidi* were often located in proximity to water (Halinen 2016: 168), such as lakes, rivers or seas, and that the sacrifices made at these sites were frequently in connection to ensuring the success of livelihood activities, like fishing or hunting (Äikäs & Salmi 2013: 66). If the scene above does indeed depict a *sieidi* offering, it would be unsurprising if the sailor was trying to calm the weather to prepare for a fishing expedition. However, in a pre-Christian context, Else Mundal (2018: 17–18) suggests that in the areas where there was close contact between the Saami and the Norse, Saami ritual performances may have influenced Norse belief systems, including the possible Norse adoption of stone cult, or even that a combined Saami–Norse culture may have developed.

Sieidi offerings were in use during the Viking Age and throughout the Middle Ages (Halinen 2016), and we know from contemporary texts that the Norse were aware of these religious spaces. In the *Eiðsivaþingslǫg*, a Christian law code from eastern Norway that survives in a manuscript from the fourteenth century, the Norse are forbidden to participate in or benefit from the following Saami ritual ceremonies:

> Engi maðr a at trua. a finna. eða fordæðor. eða a vit. eða blot. eða rot. eða þat. er til heiðins siðar hœyrir. eða læita ser þar bota.
>
> *(NGL 1: 389)*

> No man should believe in [Saami or magic practitioners], or in magical figures, or offerings, or roots, or anything which has to do with heathen practices, or seek help from these.
>
> *(Collinson 2020: 108)*

Mundal (2018: 13) suggests that there is little doubt the term *rót* in this context refers to a *sieidi*, or Saami sacred site. While it is certainly possible to interpret *rót* literally, as in the case of *Grettis saga* (ch. 79), where runes are inscribed directly onto a root, the surrounding context of the legal passage and its association with Indigenous ritual practice makes Mundal's reading more compelling. Although *sieidi* sites were most often constructed from stone, they could also be built from wooden objects (Äikäs & Salmi 2013: 66). Perhaps the law code's reference to a wooden construction suggests that the Saami people whom the Norse were seeking out for the ritual performance constructed their sites from wood rather than stone (Mundal 2018: 13). Regardless, the fact that this law code includes *sieidi* and forbids the Norse from seeking this ritual out indicates that they were in existence during the time the law code was written down, and that the Norse were aware of these ceremonies and sought them out for their own benefit (Mundal 2018: 13). As several scholars have noted, including Hermann Pálsson (1997), medieval Icelandic literature occasionally preserves cultural memories of Saami influence, particularly in its portrayal of marginal figures, magical practices, and the peripheral North. However, the, scholars engaging with the field of Indigenous history of religion inevitably reach the point of questioning how far (if at all) one can determine what reflects actual practices and beliefs and what has been altered by a Christian worldview. Although weather magic is not mentioned in connection with a *sieidi* ritual in a medieval context, perhaps it is plausible that the two could have been connected at some point. However, most of the weather-magic scenes in Old Norse saga literature are not depicted as religious ceremonies or offerings, and the terminology that is used to portray these acts emphasises that the weather-changing results stem from a magical knowledge. Trying to interpret all of these acts of weather magic as accurate reflections of a real religious practice can be problematic since these literary acts are often employed in the context of villainising both Norse and non-Norse characters, as demonstrated in the previous chapter on efficatory magic. Rather than trying to classify every example of weather-magic performance or act under one umbrella term, like 'ritual performances', 'religion', 'witchcraft' or 'magic', I would argue that evaluating each example on a case-by-case basis by considering all the contextual information available (like the terminology used, which source the example comes from, the time in which the source was written, potential audience and author biases, etc.) is more valuable. Only then can scholars determine whether a particular weather-magic episode can possibly carry some reflection of an actual practice. Scholars must also remember that the examples of magic found in Old Norse texts are not accurate portrayals of events or magical acts as they really happened, rather they are a result of Norse representations of far-removed events, as they perceived them to be from a position affected by their own worldview.

Geographic, rather than Cultural, Variations in Weather Magic

Another prevalent issue in the current field of studying medieval magic in the Nordic region is that scholars often have a tendency of grouping the Indigenous populations of Fennoscandia into one homogeneous group without considering the multiple cultural variations between communities, including the many subgroups of Saami, that are represented throughout the primary sources.[9] Saami is an overarching name that designates Indigenous peoples of Norway, Sweden, Finland and the Kola Peninsula. Saami encompasses many different cultural groups, who speak ten distinct languages (Saarikivi 2011: 78). Carl-Gösta Ojala (2009: 73) considers these variations in language to be an essential reminder that 'the Saami' are not one singular entity, neither in the past nor in the present.

Non-Norse groups other than the Saami are also represented in the surviving textual sources from the Middle Ages. For example, Bjarmians (*Bjarmar*), the inhabitants of Bjarmaland, are present in c. 30 medieval textual sources, the majority of which were written in Old Norse (Koskela Vasaru 2012: 38).[10] Mervi Koskela Vasaru (2012: 54) proposes that a close analysis of textual evidence pinpoints Bjarmaland's approximate location to the Kola Peninsula, close to the White Sea, whereas other scholars suggest that it was closer to the Northern Dvina River or Viena Karelia.[11] The Kvens from Kvenland are also present in Old Norse texts, including *Historia Norwegiæ*, *Egils saga* and *Orkneyinga saga*. Older written sources place Kvenland on the northernmost coastal landscape around the Gulf of Bothnia, while sources from the twelfth century onwards depict its approximate location to be between Hälsingjaland and Finland (Hansen & Olsen 2014: 152). Modern scholarship places the approximate location of the Kvens' settlement areas to be around the uppermost and innermost regions around the Gulf of Bothnia (Hansen & Olsen 2014: 152).

Generally, the same scholars who ignore the cultural diversity of Indigenous groups in medieval Fennoscandia also tend solely to focus on the Saami as the weather-magic experts of the north, and often ignore the other Indigenous groups that were also depicted as weather-magic practitioners during the Middle Ages. However, the contemporary magic scenes that are set in Fennoscandia are evidence that medieval scribes were aware that the Indigenous populations in this region were anything but a monoculture. The cultural variance of Indigenous weather-magic practitioners is especially evident in the weather-magic scene that transpires in the ninth book of Saxo Grammaticus' *Gesta Danorum*. In this passage, the legendary hero Ragnarr loðbrók leads his army into Northern Fennoscandia to assert political power over the local populations. During the battle scene, Saxo relays the fact that the Bjarmians use magic to bring raging thunderstorms and extreme heat to combat Ragnarr's troops:

> At Regnerus ceteris prompta sibi deditione substratis, quum quinquennem propemodum piraticam expleuisset, Biarmos nuper deuictos inualida subiectionis fide palam imperium detrectantes inuenit. Qui (Biarmos) quum aduentum eius compertum haberent, carminibus aggressi celum sollicitatas nubes ad summam usque nymborum uiolentiam impulerunt... Eosdem quoque subito remissa tempestate estuosissimi feruoris flagrantia torruit. Nec ea quidem pestis concitati frigoris magnitudine tolerabilior extitit. Itaque anceps gemine intemperantie malum uicissim affecta corpora immoderata utriusque status accessione corruptit... Ita Danorum plerique dubia celi qualitate conclusi passim oborta corporum pestilential decesserunt.
>
> *(Saxo I: 648)*

All other opponents were brought to heel after prompt surrender, but, after spending nearly five years in pirating expeditions, Ragnar discovered that the Biarmians, reduced not long before, were openly critical of his rule and he could put inadequate reliance on their passivity. When the Biarmians knew of his arrival, they addressed their magical spells to the heavens and by agitating the clouds eventually brought about huge, raging

> thunderstorms... Suddenly the tempest abated, only to be replaced by a burning, baking heat. This curse was no more endurable than intense cold. The two wicked extremes of climate following one upon the other broke the men's constitutions... So the majority of the Danes, trapped by the fluctuating sky, contracted disease, and death was widespread.
>
> *(Saxo I: 649)*

Saxo further attributes the severe weather to the Bjarmians' magical skills by explicitly stating that the storm and extreme heat was 'adulterina magis quam uera aeris' (Saxo I: 648), or 'not natural, but contrived' (Saxo I: 649).

This excerpt clearly states that the Bjarmians (*Biarmi*) are responsible for controlling the weather, and there is no mention of Saami weather manipulation. Later in this fighting episode, the Bjarmian troops are joined by a company of Saami warriors (finnar)[12] from Finnmǫrk, who are described as having impressive skills in skiing and archery. Still, Saxo includes no mention of Saami weather magic (Saxo I: 650–651). Even so, this episode has at times been treated in the secondary literature as evidence of Saami weather magic, despite Saxo's explicit attribution of the storm to the Bjarmians. Such readings are problematic because attributing the weather manipulation to the Saami ignores the many regional variations in the Indigenous communities of the north. Even Saxo himself does not present the Indigenous population of the Scandinavian peninsula as one homogenous group in *Gesta Danorum*. As Solveig Wang (2023: 40) points out, Saxo acknowledges the cultural variety in the region by specifying groups like the Bjarmians, alongside three separate Saami subgroups: the *finnar* (the expert archers represented in the scene above), who reside in Finnmǫrk; the *scritefenni*, who inhabit Norway and Sweden's eastern midlands; and finally, the *Lappia*-peoples of the Finnish and Swedish Laplands. Dismissing or ignoring the inclusion of the multiple ethnic groups that are represented in medieval texts erases a multicultural Indigenous presence in medieval spaces, which has both historical and modern consequences (Wang 2023: 40).[13] Additionally, assuming that the Saami are also the only Indigenous group that are represented as weather-magic practitioners, and that weather magic is not a culturally shared practice between the Norse and Indigenous communities, is also problematic, and reinforces the questionable tendency of scholars to 'Other' Saami characters through their use of magic. As Eleanor Barraclough (2017: 2) notes when challenging the complete separation of Indigenous and Norse cultures, many scholars writing on the literary representations and cultural perceptions of these Indigenous groups concentrate on their 'otherness', or what sets them apart from the Norse. When discussing the Saami's reputation of being able to perform weather magic, Jeremy DeAngelo claims that 'in the sagas, the depiction of the *Finnar* serves to place them in a position of the exotic Other relative to Norse conventionality' (DeAngelo 2018: 210). Barraclough (2017: 2) includes Sirpa Aalto, who describes the *Finnar* as 'an outsider group, which existed on the outskirts of Norse society' (Aalto 2003: 2) and who represent a 'negative aspect and marginal group' (Aalto 2003: 6), and John Lindow, who describes the 'Finns' as dangerous outsiders due to their skills in magic (Lindow 1995: 11), among the many leading scholars in our field who emphasise a Norse-Indigenous dichotomy, rather than acknowledging the shared cultural elements of the different cultural communities of Northern Fennoscandia.

In Old Norse literature, having knowledge in and practising magic is culturally shared by Norse and multiple Indigenous communities in Northern Fennoscandia, and weather magic is certainly no exception. There is, however, a regional variation to the type of weather magic that is performed, but the cultural ties of the weather-magic practitioners have little to no influence on when and what type of weather magic is performed. Who practises what type of weather magic seems to be determined by their geographic region and the type of landscape the magic takes place in, not their ethnicity. Naturally, Scandinavia has a large coastline and much of the population has always relied on seafaring, therefore most of the weather-magic scenes found in Old Norse sources are performed in a seafaring context. For example, most Indigenous weather-magic scenes are set around the coastal communities of Northern Fennoscandia and incorporate elements of manipulating the winds at sea. This is unsurprising, given the amount of coastline in the region. Both Saami weather-magic scenes from *Örvar-Odds saga* and *Ketils saga hœngs* that were discussed in previous chapters take place off the coast of Finnmǫrk. Similarly, in *Heimskringla*, King Óláfr Haraldsson encounters Indigenous weather magic during a harrying expedition off the coast of Finland. In *Heimskringla*, this act of weather magic is described to take place in proximity to Bálagarðssíða. Ben Waggoner (2017: 425) notes that the saga implies that Bálagarðssíða, meaning the coast of 'Bala-enclosure', is in Finland, and is normally considered by scholars to be located on the southwest coast. While Óláfr and his men are anchored for the night, the Saami summon a storm and rough weather through their magical knowledge: 'Þeir Finnar gerðu um nóttina œðiveðr með fjǫlkynni ok storm sjávar' (ÍF 2: 11) ['During the night, the [Saami] caused horrible weather and stormy seas by magic' (*Heimskringla* II: 8)]. These two portrayals of King Óláfr's encountering weather magic demonstrates the connection between Indigenous weather magic and communities close to the coast.[14] These two portrayals of King Óláfr encountering weather magic, both involving Indigenous figures and both taking place at or near the coast, demonstrate more than incidental spatial settings. The repeated association of weather magic with coastal communities suggests that the Old Norse sagas deliberately frame Indigenous magic as being tied to liminal spaces, particularly geographical edges where Norse cultural and political influence met resistance. The coast, as a site of both vulnerability and transition, becomes a narrative space where magical control over weather is most threatening. Ships are turned away, storms summoned, and Christian authority is delayed or defied at the edges of Norse control.

In both cases, the performance of weather magic by local, marginalised figures is not only a demonstration of the power of magic, but a symbolic assertion of territorial and spiritual autonomy against Christian and royal intrusion. This spatial pattern supports the wider argument that Indigenous weather magic in the sagas is strategically located at cultural and geographic frontiers, reinforcing the idea of resistance through environmental control.

However, as highlighted above, this connection between the sea and weather magic is not specific to the Indigenous communities. Regardless of whether the practitioner is associated with Norse culture, Indigenous culture or both, weather-magic scenes in Old Norse texts are often connected to a coastal setting and involve sea-weather manipulation. Throughout this book, I analyse many similar examples of Norse weather-magic practitioners summoning magic storms and coastal winds to deter attacks. Rauðr inn rammi, summoning harsh winds in *Óláfs saga Tryggvasonar* in *Heimskringla* to prevent Óláfr from reaching his homestead and demanding his conversion to Christianity, is one such example of a Norse practitioner using weather magic in a coastal context. Although Barraclough suggests that his use of weather magic 'may have been intended to align his religious world view with that of his *Finnar* allies' (Barraclough 2017: 16), there are many other examples of Norse weather magic that are not connected to Indigenous relations.

The wide use of the weather-magic-at-sea motif throughout medieval Old Norse texts, regardless of practitioner's cultural identity, is unsurprising given that the vast majority of the populations in Norway lived (and still live) close to the coast. However, these sources also reveal that Indigenous weather magic was not limited to Fennoscandia. In the fourteenth chapter of *Reykdæla saga ok Víga-Skútu*, the character called Steinfinnr performs an act of weather magic in Iceland. In this scene, Steinfinnr summons darkness and heavy winds by sitting and staring at the ground, while speaking to himself in a trance-like way. Although the text does not explicitly state that Steinfinnr comes from a Saami community, Wang (2023: 50) suggests that the combination of his name including the *-finn* component, which is often used as an indication of a character's Indigenous descent, and his knowledge of weather magic, a skill that is often tied to Indigenous groups, perhaps indicates that he has cultural affiliations with a Saami community. Although the episode takes place on the coast of north-eastern Iceland, in Bárðartjörn, it does not fall into the typical weather-magic motif where weather magic is used to help or to harm sailing expeditions. In this episode, the magical wind, alongside magical darkness, is summoned to save a bride from an unwanted marriage by sweeping her away from the ceremony.

Although Indigenous and Norse weather magic is mostly illustrated as a knowledge found in the coastal communities of Fennoscandia and Iceland, some medieval sources also demonstrate a regional variation in weather magic, where episodes of weather magic take place away from the coastline and are not used in a seafaring context. In the third chapter of *Ketils saga hœngs*, Ketill meets Gúsir, king of the Saami (*Finnakonungr*), in a forest of Finnmǫrk. Gúsir makes his appearance riding in a reindeer-sled that is surrounded by a *mjallroka mikla*, or 'wind-driven snow' (Waggoner 2012: 10). The narrative implies that a snowstorm is brought with Gúsir's arrival, which signifies that the storm is a result of magical manipulation rather than a result of natural causes.

The aforementioned offering scenes from *Hversu Noregr byggðist* and *Ynglinga saga* are also examples of weather magic that have no explicit connection to the sea. In both of these scenes, offerings are made by local communities to influence seasonal weather. The ceremonial scenes in *Ynglinga saga* take place in what is now modern-day Uppland, Sweden. While some of the Svíar communities are located on the coast, many of these communities are situated inland and rely on agricultural success more than seafaring. The weather manipulation that the Svíar community is hoping for when they sacrifice to Njǫrðr and Freyr in *Ynglinga saga* is tied to the success of agricultural harvests and the prevention of famine, rather than calming or mustering bad weather for sea journeys. Furthermore, the many medieval weather miracles described to be performed by saints in Sweden almost always relate to agriculture. These miracles will be discussed in more detail in the following chapter on weather manipulation in a Christian context.

Medieval Indigenous Wind Knots?

One type of sea-related weather magic that is commonly linked to the Indigenous communities of Northern Fennoscandia is weather knots,[15] where

FIGURE 4.1 Olaus Magnus, On Wizards and Magicians among the Finns, from *Historia de Gentibus Septentrionalibus* (1555). Image via Wikimedia Commons. Public domain.

wind would be tied into multiple knots on a rope, which would then be sold to passing sailors in their hope of ensuring good sailing weather. Cyrus Lawrence Day (1967: 45) and Ernest J. Moyne (1981: 13) incorrectly claim that the earliest attestation of wind knots in Scandinavia comes from an 'ancient' Norse myth where Vǫlundr possesses 'a long rope of bast... with knots in it at regular intervals. In each and every knot a storm wind was bound. Each week he untied a knot and freed the wind that was bound in it, and sent it south with his mad song, charged with clouds and hail' (Day 1967: 45). However, this quotation is not from an actual medieval or ancient source, but rather from Viktor Rydberg's retelling of Old Norse mythology for children, *Fädernas gudasaga*, which was published in 1887.

Many scholars claim that the first attestation of Saami weather magic is found in texts from thirteenth- and fourteenth-century England, where scholars interpret the generic northern inhabitants who sell wind knots as Saami. In his master's thesis on supernatural motifs from Northern Fennoscandia, Perabo suggests that Saami 'paganism' is mentioned in Ranulf Higden's *Polychronicon*, a fourteenth-century Latin chronicle on universal history and geography, where the Saami are 'not only shown as pagan idolaters but also as practitioners of harmful weather magic' (Perabo 2016: 83). Daria A. Trynkina also cites these sources, in addition to John of Trevisa's fourteenth-century English translations of *De proprietatibus rerum* and *Polychronicon*, as some of the earliest texts that describe the Saami's 'supernatural features' (Trynkina 2020: 104), which include the ability to create magical wind knots. R.I. Page (1962–1965) also refers to all four of these sources as the earliest references in English to 'Lapland sorcerers', even though he does acknowledge that neither the exact region nor a specific people are named in any of these sources. I disagree with Perabo, Trynkina and Page that the northern inhabitants in these texts from England reflect the Saami and their connection to weather magic with any certainty because the geographic location of the individuals that sell the wind-knot magic is incredibly vague (i.e. no specific mention of Sápmi in any of the sources, nor Fennoscandia at all, in the case of *Polychronicon*), nor do they actually explicitly reference the Saami as the makers or sellers of these magical knots. For example, both Ranulf Higden's *Polychronicon* and John of Trevisa's English translation of the same text places the wind-knots tradition on Wyntlandia, a non-specific barren island to the west of Denmark,[16] rather than Fennoscandia or Sápmi:

> Wyntlandya, þat ilond, is by west Denmark, and is a barayne lond and of men mysbyleued; þei worschippeð mawmetrie, and selleþ wynd to schipmen, þat seilleþ to hire hauenes, as it were i-closed vnder knottis of þrede; and as þe knottes beeþ vnknette, þe wynde wexeþ at her owne wille.
>
> *(Babington 1865–1866: 323)*

John of Trevisa's translation of *De proprietatibus rerum*, which is very close to the original text that was written by Bartholomaeus Anglicus (Page 1962–1965: 215), also includes a description of Wyntlandia. Rather than being an island off the coast of Denmark, the region in question is portrayed as east of the mountains in Norway:

> Vvinlandia is a countre besides the mountaynes of Norwaye towards the easte, and stretcheth vppon the clyfe of ocean: The men of that country ben strange and some that wylde and fiers: and they occupien them selfe with wytchecrafte. And so to men that saylle by their costs, and also to men that abyde with theym/for defawte of wynde, they proffer wynde to saylynge/and so they selle wynde. They vse to make a clewe of threde, and they make dyuers knottes to be knytte therin. And then they commaunde to drawe oute of the the clewe vntu three knottes or mo/or less/as they woll haue the wynde more softe or stronge. And for theyre mysbyleue fendes moue the ayre and areyse stronge tempeste or softe, as he draweth of the glewe more of lesse knottes.
>
> *(Page 1962–1965: 215)*

This description of the geographic location of Wyntlandia could perhaps indicate that it falls within the region of modern-day Lapland, as Page suggests (1962–1965: 224), but without a clear reference to the cultural affiliation the practitioners, it is impossible to assert that this is an early attestation of Saami or other Indigenous weather magic.

The earliest known appearance of the wind-knot motif in a Scandinavian source survives in Olaus Magnus's *A Description of the Northern Peoples*, published in 1555.

Here, Magnus describes

> a time when the Finns, among other pagan delusions, would offer wind for sale to traders who were detained on their coasts by offshore gales, and when payment had been brought would give them in return three knots tied in a strap not likely to break.
>
> *(HGS.3.16: 173)*

Unlike the previous sources from England, Olaus Magnus places the use of weather knots directly in Northern Fennoscandia and attributes the magical practice to the 'Finns'. Although *Historia de gentibus septentrionalibus* (1555) was one of the most influential texts among European audiences at the time, Liv Helene Willumsen (2013: 358) reminds us that it is important to note that his account of the Indigenous populations, as well as other Far Northern peoples, is one from an outsider's perspective. Magnus never actually travelled to Northern Fennoscandia himself and relied on others' descriptions of these people and cultures when writing his text. It is therefore implausible to link his account of wind knots to actual practices or beliefs circulating in Northern Fennoscandia itself, or to use it as evidence that the tradition stemmed from an earlier, medieval origin in the region.

At the same time, Olaus Magnus's account of Saami wind-selling offers a striking example of how medieval tropes were reconfigured in the early modern imagination. The motif of knotted ropes resurfaces in his text not as a practical detail of ritual performance but as part of a broader catalogue of marvels designed to intrigue a European readership. For Olaus Magnus, the Saami appear less as historical agents than as embodiments of a liminal northern landscape where dangerous weather and dangerous people coincided. His description is therefore significant not because it provides reliable ethnography, but because it illustrates how the idea of Saami wind-magic was shaped, exaggerated, and disseminated in sixteenth-century Europe.

The fact that Olaus's description reworks motifs also found in medieval sources underscores the importance of considering cultural continuities across period boundaries. While this book focuses on medieval evidence, it is clear that wind-magic motifs continued to circulate and evolve well into the early modern period. Early modern scholarship has highlighted this trajectory in valuable ways: Liv Helene Willumsen (2013) and Rune Blix Hagen (2017) have shown how weather magic surfaced in Finnmark witchcraft trials, and Louise Nyholm Kallestrup (2025) has traced its role within a Danish context. Engaging briefly with this scholarship signals that the motifs preserved in medieval Old Norse sagas did not disappear with the Middle Ages but were reshaped within new legal, theological, and folkloric frameworks. At the same time, the differences remain substantial: medieval texts present weather magic as part of everyday narrative and customary practice, whereas early modern sources increasingly frame it as evidence in judicial proceedings and as a subject of demonological theory.

Learning Weather Magic from the Saami

Hansen and Olsen (2014: 50) highlight that it is significant how the Old Norse sagas portray the Saami as having close relationships with the Norse. Many of these texts emphasise the Saami's knowledge of magic, and Saami characters are often cast as healers, advisors, and teachers of magic for Norse characters (Hansen & Olsen 2014: 50). As pointed out earlier in this chapter, the inclusion of Saami rituals and prophecies in the *Eiðsivaþingslǫg*, alongside the forbiddance of Norwegians seeking out the Saami's help through their performance non-Christian ceremonies, indicates that these religious practices stem from a tradition that was well established by the time of Norway's conversion to Christianity, and that Norse Christians still sought Saami guidance at the time that the extant manuscripts were written (Mundal 2018: 13).

Solveig Marie Wang (2023: 42) suggests that the legal issue of seeking Saami help and teaching lies in the practice of non-Christian ceremonies, and that Christians sought their Saami neighbours out for non-Christian services, not that the practitioners' ethnicity was problematic (i.e. their Saami identity). Furthermore, we cannot infer that the specific mention of the Saami in these legal texts implies that the Norse themselves were not also conducting non-Christian practices at the time. Else Mundal (2018: 18) also emphasises that it is important to remember that the eastern Norwegian law codes do not specify that the Saami performing non-Christian rituals were non-Christians themselves, an omission that scholarship tends to overlook. As mentioned earlier, she (Mundal 2018: 18) reminds scholars that Saami conversion did not happen overnight following the sixteenth-century Reformation; rather, it started around the same time of Norse conversion and was a long, slow process. However, in these legal codes, the practices themselves are designated as *heiðinn siðr*, or 'heathen customs', and the Saami are referred to as *fordæða*, a highly pejorative term in the Old Norse corpus typically applied to those considered evil-doers or morally corrupt figures. While not all uses of *fordæða* refer directly to magic, the term frequently appears in contexts involving outlawry, non-Christian belief, and magical harm. Even if the Saami people in question are not explicitly called non-Christians in the legal sources, church authorities clearly took issue with their continued practice of non-Christian rituals and ceremonies, as seen through the forbiddance of seeking these ceremonies out. Regardless, the eastern Norwegian law codes do suggest that Saami rituals and practices did influence Norse communities to the extent that they were sought out enough for them to warrant an appearance in the texts. By the thirteenth century, the Norse seeking the Saami's magical skills was a common motif, which is demonstrated when Saxo Grammaticus portrays Óðinn seeking Saami magic (Wang 2023: 77). As Hansen and Olsen (2014: 50) point out, Norse characters learning magic from their Saami neighbours is also a common motif that is often reflected in Old Norse sagas. For example, in Chapter 19 of *Ǫrvar-Odds saga*, a young Ǫgmundr is sent to live with the Saami in Finnmǫrk to learn '*galdra ok gerninga*' (Boer 1888: 126), or magical spells and acts. This recurrent theme of acquiring magical expertise from the Saami underlines their perceived authority in the realm of magical knowledge within Norse society. While the specific motivations for these journeys often involved seeking prophecy (*spyrja spá*), the broader association of Saami ritual specialists with the knowledge of varied forms of magic, including weather manipulation, suggests that such knowledge may have been among the skills attributed to them. This reinforces the intertwined nature of Saami and Norse cultures in the context of magical belief and performance. It is therefore unsurprising that the Norse would have also been portrayed as learning weather magic from their Indigenous neighbours. This transferal of knowledge is portrayed in *Fagrskinna*. In this text, Gunnhildr *konungamóðir* is first introduced during one of Eiríkr blóðøx's raids in Finnmǫrk and Bjarmaland, where he sees his future wife for the first time. Here, it is explained that Gunnhildr is being fostered and educated by Mǫttul, king of the Saami. Mǫttul is described as very skilled in magic and the text implies that he passes his magical skills onto Gunnhildr during her period of fosterage. Later in the text, Gunnhildr is accused of summoning a treacherous storm at sea that almost drowns Hákon goði and his men. Lyonel Perabo (2016: 124) suggests that her ability to practise weather magic stems from her time living with Mǫttul, as weather magic is a common skill that is associated with the Indigenous cultures of Fennoscandia. This recurrent theme of acquiring magical expertise from the Saami underlines the significant influence and high regard of Saami knowledge of weather magic within Norse society, further illustrating the intertwined nature of these two cultures in the context of magical performances and beliefs.

Conclusions

The study of weather magic as a motif in Old Norse saga literature, alongside relevant legal documents that contribute to the understanding of the contemporary Norse attitudes towards Indigenous magic, reveals a cultural exchange and mutual influence between the Norse and Saami communities. By examining the multitude of weather-magic depictions in Old Norse literature, performed both by Indigenous and Norse characters, this chapter reveals the complexity and significance of weather magic in shaping the shared cultural and religious landscape of medieval Fennoscandia.

The Old Norse sagas frequently depict Indigenous characters as possessing extensive knowledge of weather magic, positioning them as influential figures within Norse narratives. This relationship is vividly illustrated in the sagas, where Indigenous individuals often act as advisors, healers, and teachers to Norse characters. For instance, in *Fagrskinna*, Gunnhildr konungamóðir is fostered and educated by Mǫttul, the Saami king, who is described as highly skilled in magic. This fosterage not only underscores the close relationship between the Saami and Norse but also highlights the transfer of weather-magic knowledge from the Saami to Norse individuals. Gunnhildr later uses her acquired skills to summon a treacherous storm at sea, demonstrating the practical application of this knowledge through a motif that is frequently depicted in Old Norse literature as being performed by both Indigenous and Norse characters.

The motif of Norse characters seeking Saami magical skills is a recurrent theme in Old Norse literature. This cultural exchange is emphasised in several texts, reinforcing the notion that Saami knowledge of weather magic was highly esteemed. In chapter 19 of *Ǫrvar-Odds saga*, young Ǫgmundr is sent to live with the Saami in Finnmǫrk to learn *galdra ok gerninga* (magical spells and acts). This narrative clearly illustrates the intentional and formal transfer of magical knowledge, including weather magic, from the Saami to the Norse. Such examples underscore the depth of interaction and mutual respect between these communities in the realm of magical practices, as it is portrayed in a literary context.

Weather magic is depicted as a shared practice among Norse and Indigenous groups, with a strong emphasis on its application in coastal and seafaring contexts. This is evident in the frequent situating of weather-magic scenes in coastal settings within Old Norse texts. For example, in *Ketils saga hœngs*, a coastal Indigenous community calls a storm that affects Ketill's sailing course, leading him to meet and eventually defeat king Gúsir. This scene highlights the practical use of weather magic in maritime contexts, reflecting the importance of such knowledge for seafaring communities. Similarly, in *Heimskringla*, King Óláfr Haraldsson encounters Indigenous weather magic during an expedition off the coast of Finland, where the local Indigenous community summons a storm to hinder his progress. These literary examples illustrate how weather magic was portrayed as intertwined with the daily lives and survival of coastal communities.

The chapter also challenges the homogenisation of Indigenous groups, emphasising the cultural diversity within Saami communities and other Indigenous populations of Fennoscandia, such as the Bjarmians. This diversity is crucial for understanding the different depictions and uses of weather magic across various groups. In *Gesta Danorum*, Saxo Grammaticus describes the Bjarmians using magic to bring violent thunderstorms and extreme heat to combat Ragnar loðbrok's troops. This scene clearly differentiates the Bjarmians' magical contributions to the battle from the non-magical contributions of the Saami, highlighting the contemporary awareness of the multiple Indigenous groups in the region. Such depictions underscore the importance of recognising the unique cultural identities and practices of different Indigenous communities that existed in Scandinavia during the medieval period.

Recent scholarship calls for recognising the cultural fluidity in medieval Fennoscandia, challenging the perception of a strictly monocultural Norse society distinct from what has previously been delineated as the Indigenous 'Others'. This is evident in the mutual influence between the Norse and Saami, particularly in the context of weather magic. The development of a combined Saami–Norse cultural milieu is suggested in areas where close contact and mutual influence occurred. For instance, the use of weather magic by Gunnhildr, as described in *Fagrskinna*, indicates the integration of Saami magical practices into Norse society. This example highlights how cultural boundaries were fluid, allowing for significant intercultural exchange and the blending of traditions.

The study of weather magic from Norse-written sources presents challenges due to the lack of direct Indigenous perspectives. The chapter emphasises the need for careful, context-sensitive analysis to avoid exoticising or misrepresenting Indigenous practices. For example, the description of Saami weather magic in early modern sources like the witch trial records of Quiwe Baarsen and Anders Poulsen, and the comparison of these trials to the medieval period, must be approached with extreme caution. These records, produced under traumatic conditions, cannot be taken at face value but require critical examination to understand the context and biases involved. This methodological rigour ensures a more accurate and respectful representation of Indigenous weather-magic practices, as they were understood in the medieval period.

The chapter advocates for evaluating each episode of weather magic found in Old Norse texts individually, considering all contextual factors to determine its authenticity and significance. This approach is necessary to avoid oversimplification and to appreciate the nuances of each example. For instance, the weather-magic scene in *Sturlaugs saga*, where a magic practitioner bargains for favourable winds, differs from religious offering scenes in other texts. By analysing the specific context and details of each instance, scholars can better understand the diverse applications and meanings of weather magic in medieval Scandinavia.

In conclusion, the exploration of weather magic in medieval Scandinavian texts reveals the complexity of cultural exchange, mutual influence, and shared knowledge between the Norse and Indigenous communities. The Indigenous communities of Northern Fennoscandia, being portrayed with extensive knowledge and mastery of weather magic, played a crucial role in shaping Norse magical narratives. The frequent depiction of Indigenous weather magic in Old Norse sagas and other texts underscores the high regard for their expertise and the significant intercultural exchange that characterised medieval Fennoscandia. By examining specific examples and considering the diverse contexts in which weather magic was employed, this study challenges simplistic dichotomies and highlights the rich, shared cultural heritage of these communities. This approach not only deepens our understanding of medieval weather magic as a cultural concept in medieval Scandinavia but also underscores the importance of recognising the interconnectedness and fluidity of cultural boundaries in the historical study of magic.

Notes

1 See the chapter on weather-magic performance for a discussion on the lack of physical evidence left by Norse or Indigenous weather magic.

2 The term *noaidi* is from the North Saami language. Wang (2023: 94) notes that the word *noaidi* is first documented in late-early modern sources and she therefore cautions scholars against using this name as a descriptor for the ritual performers portrayed in medieval texts.

3 This trial was introduced and discussed in the introduction of the present study.
4 For a detailed overview and analysis of Quiwe Baarsen's and Poala-Ánde's trials, see Hagen 2014.
5 A fellow inmate murdered Poala-Ánde before the court reached a sentence.
6 For example, Niemi (1993) and Hagen (2002).
7 For a recent overview of these trials and their political and theological contexts, see Kallestrup 2025.
8 *Sieidi* sites seem to be more common in the northern and central Saami communities, whereas they are mainly clustered around Semisjaur-Njárg and Granbyn on the Swedish side (Hansen & Olsen 2014: 219). The term *sieidi* is from the North Saami language.
9 For examples, see Moyne 1981, Lionarons 1997, and DeAngelo 2010, 2018.
10 There are a few surviving references to the Bjarmians in Latin texts, like Saxo Grammaticus' *Gesta Danorum*, alongside one surviving reference from the Old English Ohthere's account (Koskela Vasaru 2012: 38).
11 See Koskela Vasaru 2012 for a summary of each theory regarding the possible location of Bjarmaland, beginning with arguments from the sixteenth century onwards.
12 Finnr (singular) and Finnar (plural) are the Old Norse terms for the people who are interpreted as the ancestors of the modern Saami (Ojala 2009: 84; Mundal 1996: 98; Barraclough 2020: 71). Saxo uses the term 'Finni'.
13 Wang (2023: 216) notes that Saami history and their representation, or perceived lack thereof, in medieval sources is directly tied to contemporary conflicts, as medieval sources have often been used in modern political discourses and lawsuits to disadvantage Saami people (e.g. to either argue for or against Saami rights to certain land and water spaces).
14 It is, however, interesting to note that the earlier portrayals of this scene, found in *Fagrskinna* and *Helgisaga Óláfs Haraldssonar*, do not include the use of weather magic by local groups to combat the king's attacks.
15 See chapter 3 of the present study for more discussions regarding the weather knot motif.
16 Many scholars, as described above, place the island in Scandinavia. DeAngelo (2018: 213), however, suggests that the location in question is the Isle of Man.

5

MAGIC OR MIRACLE?

Christian Counter-Magic and Weather Saints

The distinction between acts of magic and miracles that are depicted in medieval Scandinavian literature has been a subject of considerable scholarly debate (for example, McCreesh 2018, 2006; Mitchell 2015, 2011; Meylan 2011; Raudvere 2002). Central to this discourse is the exploration of how these phenomena are represented in the Old Norse corpus and the broader context of medieval European literature. As Mitchell (2011: 51) suggests, the outcomes of medieval magic and miracles in these texts often appear similar, particularly in their ability to alleviate the day-to-day stresses of life. This chapter seeks to dive deeper into these similarities and differences by examining a selection of episodes of weather miracles and weather magic that are set in Scandinavia, found in continental and Old Norse texts. The main aim of this chapter is to investigate how medieval Scandinavians and those who wrote about the region in the Middle Ages separated the cultural concepts of weather miracles and weather magic, through the way that these acts are represented in medieval texts. By doing so, this chapter also aims to uncover whether the literary depictions of Christian weather miracles in medieval Scandinavia were influenced by non-Christian weather magic, and to what extent these practices were integrated or reinterpreted in a Christian context.

As demonstrated in the previous chapters of this study, weather manipulation, whether conducted through magic or miracles, played a significant role in the daily lives of medieval Scandinavian communities. This importance is reflected in Old Norse sagas, Old Norse poetry, and other medieval texts, where control over the weather is often portrayed as a means to overcome natural hardships, such as droughts, storms, and harsh winters. The reliance on favourable weather for agricultural success and safe sea voyages

DOI: 10.4324/9781003746058-8

underscores the critical need for weather manipulation practices. This chapter will explore how these needs were addressed in literary scenes through both non-Christian and Christian practices, revealing a nuanced distinction between magic and miracle that medieval Scandinavians that is reflected in the surviving literature.

The primary distinction between weather magic and weather miracles often lies in their perceived source of power and the cultural framework within which they are practised. Non-Christian weather magic is typically depicted as deriving from the innate power of individuals, such as magic practitioners designated as *fjǫlkunnigr, margkunnigr* or *spákona*, who use their special knowledge to manipulate and control the elements. In contrast, Scandinavian weather miracles are portrayed as acts of divine intervention, facilitated by saints, bishops, or other holy figures who channel the power of God to achieve similar outcomes. This dichotomy, however, is not always clear-cut, and there are numerous instances where the lines between magic and miracle blur when portrayed in Old Norse literature, suggesting a more nuanced understanding of these practices in medieval Scandinavia.

Weather Magic or Weather Miracles?

Michael E. Goodich (2007) explores the theology of medieval miracles and how medieval theologians distinguished acts of *miracula* (acts that were the result of God's direct influence) and *mirabilia* (natural occurrences without divine intervention, albeit rare). In this study, I will be defining miracles as an event described in the Old Norse corpus that exceeds the powers of nature and is portrayed as a result of God's divine intervention. When examining the relationship between magic and miracles in thirteenth-century English miracle collections, Valerie I. J. Flint underlines 'the very closeness of magic and miracles, for both seek to procure results which transcend nature, and to do this through the medium of a human practitioner' (Flint 2002: 117). This same connection between magic and miracles is seen in medieval Old Norse sources, where both acts of magic and miracles are performed by human practitioners, with the aim to control certain scenarios that humans cannot normally control, like the weather. This chapter will analyse the difference between episodes of weather miracles and weather magic depicted in the Old Norse corpus, and demonstrate possible influences that non-Christian weather magic may have had on Norse weather miracles.

The difference between magic and the miraculous within the Old Norse corpus is contentious within the field of medieval Scandinavian studies. For example, Meylan suggests that 'from the perspective of its rites, representations and agents, the magic we find in the *Íslendingasǫgur*, royal biographies, the Eddas, or Scandinavian law codes has little to do with the Christian magic as it appears in the apostles' or bishops' sagas, and more generally in continental

European ecclesiastical thought' (Meylan 2011: 113). However, this chapter aims to demonstrate that in the case of weather manipulation, there is not a clear division between weather-magic performances and weather miracles in the acts themselves.

Contrary to Meylan, Stephen Mitchell (2011: 11) proposes that it is difficult to differentiate between the magic that elite or learned members of society used, which they would call 'miracles', over the magic that was being used by those who the elite classified as their social inferiors, like those who would be categorised as magic practitioners using Old Norse designations like *fjǫlkunnigr, margkunnigr, spákona*, etc. However similar magic and miracles are, Mitchell (2011: 11) does point out that in the minds of many medieval church authorities, there were acute differences between magic and miracles, which were divided by the source of power that made the outcome effective. He notes:

> a truism, as appropriate to the medieval way of thinking as it is of much modern scholarship, is that practitioners of magic... manipulate that power, command it, ordering the source of power to do the performer's will, whereas the religious person comes as a supplicant, praying to the source of power.
>
> *(Mitchell 2011: 11)*

However, as previously demonstrated in this study, there are many cases where individuals and communities plead to the Old Norse deities to manipulate their natural surroundings (this will be explored further later in this chapter). Flint (2002: 123) brings up an inevitability in the similarity between acts of magic and miracles in the Middle Ages, simply because the patrons of medieval magic performers and miracle workers often sought the same outcomes, like cure from illness, aid in love affairs, and changing the weather in sailing and agricultural contexts.

When considering the corpus of Christian miracles that are depicted throughout Old Norse saga literature, it is unsurprising that weather miracles are amongst these miraculous acts. The notion that God has control over the weather was widespread throughout Christian Europe (Loomis 1948: 45), and medieval Iceland was no exception (McCreesh 2018: 24).[1] This belief is reflected in Old Norse texts, where individuals would call to God for a change in the weather. In *Guðmundar saga dyra* (Ch. 14), a vow to God changes the direction of the wind to save Guðmundr's farmstead from a fire. Although the example does not directly attribute the change in weather to divine intervention, the narrative implies that God answered the prayers for wind (McCreesh 2018: 25). In Old Norse texts, God is not the only divine Christian figure portrayed to manipulate the weather on behalf of Christian believers. Throughout these texts, saints and bishops were portrayed as having the ability to change the weather, often through a direct appeal to God. In *Þorláks saga byskups*, Bishop Þorlákr intercedes on behalf of sailors in distress by calming multiple sea storms (Þorláks: 27–28).

Members of the medieval Church and religious scholarly community did not always deny the existence of non-Christian magical practices but rather, insisted that God yielded a greater power than the humans who performed these magical acts (Raudvere 2002: 82). By the late-fifteenth century, it was an accepted view that weather magicians existed. In the infamous *Malleus Maleficarum* (Hammer of Witches), Heinrich Kramer wrote that magic practitioners could summon storms of hail, thunder, and lightning (Kramer 2009: 109). However, scholarly members of the medieval church did not universally accept or deny the existence of non-Christian weather magic during the Middle Ages. In *De Grandine et Tonitruis* (c. 815), Agobard of Lyon (1981) clearly states that humans were not able to control the weather by any means. Additionally, as highlighted in the introduction and second chapter of this book, Pope Gregory VII wrote a letter to King Haraldr III of Denmark in 1088, reinforcing Agobard's statement by writing that only God has the ability to control the weather (Emerton 1967: 153), not human practitioners of magic. Regardless of whether certain medieval theologians acknowledged that non-Christian humans were able to alter the weather via magical acts or by channelling the powers of demons, many did consistently acknowledge that the manipulation of the weather was achievable through weather miracles performed by saints, the difference in magic and miracle being that these religious individuals were pleading with God, who would then perform the weather manipulation act on their behalf, rather than having the abilty to manipulate that power on their own accord. Although a change in the weather is the same end goal for both acts of non-Christian weather magic and Christian weather miracles, the performer and their religious association is also what differentiates the medieval concept of weather magic and weather miracle. This remainder of this chapter will explore the differences in how non-Christian weather magic and Christian weather miracles are portrayed in the Old Norse corpus, and whether non-Christian beliefs or practices may have influenced the depiction of weather miracles in medieval Scandinavia.

The morality of practising weather magic was complicated in medieval Scandinavia, in that a direct manipulation of the weather was not always portrayed as a condemnable act of magic, nor was it restricted to non-Christian practice. In Old Norse saga literature, weather manipulation is attributed to non-Christians and Christians individuals alike. Alongside the many non-Christian magic practitioners that have been previously highlighted in this study, Christian bishops and saints are also often portrayed as having the ability to change the weather, albeit through prayer rather than the emic-Norse magic, like *galdr* or *seiðr*. In the *biskupasǫgur*, Bishop Jón Ǫgmundarson uses prayer twice to change the weather to save a community from drought and prolonged winter (BS II: 49; BS II: 50–51), and Bishop Guðmundr Arason prays for favourable winds during his sea-voyage home (BS II: 249–250). Although Christian saints and non-Christian magic practitioners are portrayed as having a similar ability to control the weather in Old Norse texts, McCreesh (2006: 675) suggests that Christian and non-Christian weather manipulation are normally represented as diametrically opposed in the *Íslendingasǫgur*. McCreesh explains her suggested opposition with the following three points: firstly, non-Christian weather magic is deliberately summoned with the use of intentional magical acts, like chants or spells, whereas she suggests that weather miracles are portrayed without direct human influence. However, I disagree with this separation, as I have previously noted several examples where human prayers to God or appeals to a Christian saint are used to change the weather. In the scene from *Vatnsdœla saga* (Ch. 34) that I previously analysed in the chapter on weather-magic terminology, Bárðr stirfinn is depicted as a Christian who calms a blizzard that was brought by non-Christian weather magic. In this scene, Bárðr and the rest of the Christian community who partake in the count-spell have a direct influence on the change in weather.

McCreesh's second point to propose a separation between non-Christian and Christian weather manipulation is that non-Christian weather magic is often employed in the *Íslendingasǫgur* with a motivation of 'evil-doing' (McCreesh 2006: 675). Yet, this is certainly not the case in either of the previously discussed episodes from *Vatnsdœla saga* (Ch. 34) and *Finnboga saga* (Ch. 34). In both of these episodes that portray the same event in slightly different ways, the motivation for Helga or Dalla to summon a blizzard, rumoured or not, is to prevent a deadly duel. The incentive in both of these representations of weather magic is to save Bergr from a duel, which stems from the desire to protect a loved one, rather than an act of maliciousness. Furthermore, as demonstrated in the previous chapter on efficatory magic, the categorisation of non-Christian weather magic into harmful and beneficial categories is not as straightforward as suggested here. Although the repercussions of a magical act could superficially suggest that the motivation of using weather magic was done with 'evil' intent, such as to cause harm to an innocent person, the actual motivations for using weather magic is rarely clearly outlined within the majority of the Old Norse examples. A lack of description surrounding the use of weather magic often obscures the practitioner's motivations, as seen in the previously discussed scene from *Ketils saga hœngs* (Ch. 3), where the local Indigenous community summons a storm against Ketill's boat. As I have previously highlighted in the chapter on efficatory magic, although the Indigenous community's motivation for employing weather magic is omitted from the text, it is likely that their magical act is used to protect the community's livelihood. Although saga writers rarely provide the view point or outline the motivation of magic practitioners, particularly those of marginal characters who perform weather-magic acts, these characters do not employ magic in these texts without a specific reason that the audience could interpret as justified to some extent. I would argue that there are no examples of non-Christian weather magic in the *Íslendingasǫgur*, or within the Old Norse saga literature corpus, that are presented as having been committed purely with the motivation of 'evil-doing', as each individual that practised weather magic is portrayed as using magic as a means of protecting themselves, their family, or their wider community.

Perhaps the difference between non-Christian and Christian weather magic could be that the surrounding context of non-Christian weather-magic performances in the Old Norse sagas evokes a sense of disapproval in the audience, by intentionally using terminology that suggests the magical acts are reprehensible,[2] whereas acts of Christian weather magic are portrayed more positively. Yet, once again, not every example of non-Christian weather magic is presented as morally wrong within the text. Þrándr's use of weather magic in the previously discussed scene from *Færeyinga saga* is presented by the saga narrator as potentially justified, in that it is a means to resist political and religious colonisation. Each of the weather-magic acts highlighted in the chapter on efficatory magic demonstrates that it is impossible for morality to be the dividing factor between an act of non-Christian weather magic and an act of Christian weather magic.

McCreesh's (2006: 675) final point put forth in support of the suggested dichotomy between non-Christian and Christian weather magic is that weather summoned by non-Christian magic practitioners is often stormy, whereas Christian weather magic normally produces a favourable change in the weather and benefits the local community. This is perhaps true of many of the depictions of non-Christian weather magic found in the *Íslendingasǫgur*, but the Old Norse corpus certainly contains a variety of examples where non-Christian weather magic is employed to create a favourable change in the weather to benefit others, rather than only benefiting the practitioner. A comparison between Saint Þorlákr's and Óðinn's ability to calm storms at sea plainly demonstrates that both non-Christian and Christian weather magic was depicted as a way to bring favourable weather in rough sailing conditions. Although Óðinn's and Þorlákr's skills in weather manipulation seem to overlap, Þorlákr's weather manipulation is deemed miraculous, whereas Óðinn's ability to alter the weather is attributed to his magical skill set. As Mitchell (2011: 49) highlights, the division between miracles and magic often relies on the viewpoint of the observer and the observed. From a comparison of scenes depicting either Þorlákr or Óðinn, the division between weather magic and weather miracles is no different. Take, for example, the previously discussed scene from *Vǫlsunga saga* where Óðinn rescues Sigurðr and his crew from a storm.[3] If one were to replace Óðinn, a non-Christian god, with Saint Þorlákr, the audience would perhaps draw different conclusions about the nature of the event and the ones seeking help. When sailors call Saint Þorlákr to calm a storm at sea, as is depicted in *Þorláks saga byskups*, it is portrayed as a miracle. Yet when a non-Christian god manipulates the weather in a very similar context, the text represents the act under the realm of magic. The main difference between the contexts of the two episodes, outside of who is causing the weather manipulation, is that in *Þorláks saga byskups*, the men appeal directly to Bishop Þorlákr, while in *Vǫlsunga saga*, Óðinn materialises on his own agenda.

Comparing episodes of Christian and non-Christian weather manipulation in the *Íslendingasǫgur*, and those found in the wider Old Norse corpus, demonstrates that the separation of the miraculous and magic is far more nuanced in medieval Scandinavia than McCreesh's suggestion that these episodes of weather magic and miracles sit in diametric opposition to one another. The morality of practising weather magic in medieval Scandinavia is complicated, and the specific motivations, malicious or not, in which these acts are performed in the *Íslendingasǫgur* are hardly given. It is therefore impossible to categorise every episode of weather-magic performance as 'wrong', in opposition to the 'rightness' of miraculous weather magic.

Individual- or Community-Driven?

Another intriguing aspect of weather manipulation in Old Norse literature is the potential distinction between weather miracles and weather magic based on whom the magic is intended to benefit. While weather miracles often appear in texts to benefit the wider community, weather magic is frequently depicted in the Old Norse sagas as serving the interests of the individual practitioners. The portrayal of weather miracles in Old Norse sagas such as the *biskupasǫgur* suggests that weather miracles were often conducted with the aim of benefiting the collective welfare of the community. For instance, Bishop Jón Ǫgmundarson's aforementioned prayers to end drought and prolonged winter not only save individuals from hardship but also safeguard the agricultural prosperity of the entire community. Similarly, Bishop Guðmundr Arason's pleas to God for favourable winds during his sea voyage home ensures the safety and well-being of all on board the ship.

On the other hand, instances of weather magic found in Old Norse literature are often depicted as magical means to serve the personal interests or vendettas of individual practitioners. An example is found in the previously discussed scene from *Gísla saga Súrssonar*,[4] where Auðbjorg summons a snowstorm to exact vengeance upon her son's assailants:

> Kerling fær ekki sofnat um nóttina, svá var henni bimbult. Veðr var kalt úti ok logn ok heiðríkt. Hon gengr nǫkkurum sinnum andsœlis um húsin ok viðrar í allar ættir ok setr upp nasarnar. En við þessa hennar meðferð

> þá tók veðrit at skipask, ok gerir á fjúk mikizt ok eptir þat þey, ok brestr flóð í hlíðinni, ok hleypr snæskriða á bœ Bergs, ok fá þar tólf menn bana.
> (*ÍF* 6: 59–60)

> Old Auðbjorg was so uneasy that she had no sleep that night. It was cold outside, but the air was still and the sky cloudless. She walked several times widdershins around the outside of the house, sniffing in all directions. As she did this, the weather broke and a heavy, blustering snowstorm started up. This was followed by a thaw in which a flood of water gushed down the hillside and sent an avalanche of snow crashing into Berg's farmhouse. It killed twelve men.
> (*CSI* 2: 22)

In this narrative, the magic serves the narrow, personal agenda of Auðbjorg rather than benefitting the broader community.

While the distinction between community-oriented weather miracles and self-serving weather magic holds true in many cases, there are notable exceptions and nuances to consider. For example, as we have previously seen in *Vatnsdœla saga*, Bárðr stirfinn calms a blizzard using weather magic, benefiting his entire community. However, it is notable that Bárðr and his fellow practitioners are depicted as Christian, which perhaps in the eyes of the saga narrator and audience aligns his motives with the saints who perform Christian weather miracles, rather than those who practise non-Christian weather magic. Furthermore, instances of weather magic benefitting communities rather than individuals also exist. In *Ketils saga hængs*, as demonstrated previously, the local Indigenous community summons a storm to protect their livelihood and communal interests. Like the moral ambiguity surrounding performances of weather magic as it is depicted in Old Norse literature that is demonstrated in the third chapter of this study, the division of weather magic and weather miracles based on the motives is not always straightforward in these texts, where motivations and outcomes of weather-magic events may vary depending on cultural, religious, and individual factors. That being said, all instances of weather miracles that are depicted in Old Norse texts stem are performed to benefit a community, rather than the miracle benefiting an individual.

Using Weather Miracles against non-Christian Weather Magic

Regardless of the difference between ways in which non-Christian weather magic and Christian weather magic are portrayed in medieval Icelandic literature, the belief that God, or other divine individuals within Christianity, could reverse or overcome any form of non-Christian weather magic is consistent in Old Norse saga literature, with particular popularity in the *konungasǫgur*. Throughout the kings' sagas, weather magic is regularly overturned by Christian individuals, normally bishops or kings, who would use their faith in God to overcome non-Christian weather magic. This is demonstrated, for instance, in the eightieth chapter of *Ólafs saga Tryggvasonar*. After King Oláfr's multiple attempts to enter the fjord in Sálpti are blocked by Rauðr inn rammi's weather magic, Bishop Sigurðr uses prayer and holy water to calm the tumultuous weather long enough for Oláfr's ships to pass through the fjord. The saga episode implies that by channelling the power of God through prayer and ritual, Bishop Sigurðr was able to overcome non-Christian magic.

The actual performance that Bishop Sigurðr uses to still the storm is described in a surprising amount of detail, especially in comparison to how the performance of non-Christian weather-magic is normally portrayed in Old Norse textual sources:

> Sigurðr byskup tók allan messukrúða sinn ok gekk fram í stafn á konungsskipi, lét tendra kerti ok bar reykelsi, setti róðukross upp í stafninn, las þar guðspjall ok margar bœnir aðrar støkkði vígðu vatni um allt skipit. Síðan bað hann taka af tjǫldin ok róa inn á fjǫrðinn. Konungr lét þá kalla til annarra skipa, at allir skyldi róa eptir honum. En er róðr var greiddr á Trǫnunni, þá gekk hon inn á fjǫrðinn, ok kenndu þeir engan vind á sér, er á því skipi røru, ok svá stóð toptin eptir í varrsímanum, at þar var logn, en svá laus sjárokan brot frá hvárntveggja veg, at hverti sá fjǫllin fyrir.
> *(ÍF 26: 326)*

> Bishop Sigurðr got out all his Mass vestments and went forward to the prow on the king's ship, had candles lit and took incense, set a crucifix up on the prow, read from the Gospel and many other prayers, sprinkled holy water all over the ship. Then he bade them take down the awnings and row into the fiord. The king then had someone shout to the other ships that everyone was to row behind him. And when the oars were plied on Tranan, then she went into the fiord, and those who were rowing on that ship felt no wind on themselves, and the walls of the wake behind them stood up so it was calm in there, but there was such widespread spray on each side of them that because of it the mountains could nowhere be seen.
> *(Heimskringla I: 203–204)*

Here, the audience is made acutely aware of each step that the bishop uses to overcome Rauðr's non-Christian weather magic. Bishop Sigurðr employs a combination of spoken words (i.e. prayers and the reading of the Gospels) and religious symbols (i.e. incense, candles, holy water, and a crucifix) to combat the magical winds. The episode's inclusion of the vocal element of reading from the Gospel and saying many other prayers out loud is perhaps reminiscent of how the vocal performance of non-Christian weather magic is emphasised throughout the Old Norse corpus, as demonstrated in the previous chapter on weather-magic performance; however, the emphasis in this scene tends to be more focused on the bishop's physical action.

These examples of Christian authorities overturning non-Christian magic are employed to demonstrate that God's power is mightier than the weather magic that is performed by non-Christians. The reversal of non-Christian weather magic is, however, not the main function of Christian weather miracles. Although there are scenes in the Old Norse corpus where weather miracles are used as a means to demonstrate the power of God in comparison to a non-Christian power, the majority of weather miracles scenes focus the manipulation of weather to bring an end to hardship, like a famine or storm at sea.

Cultural Loan Shifts

One example that foreshadows the syncretism explored in this section is the enduring ritual formula *til árs ok friðar*, meaning 'for a good harvest and peace'. While the phrase does not explicitly mention weather, the term *ár* denotes agricultural prosperity and seasonal abundance, which are both outcomes deeply dependent on favourable weather conditions in medieval Scandinavia. This expression appears in both pre-Christian and Christian sources, including legal codes and religious supplications, and reflects a cultural continuity in how communities voiced their desires for environmental stability.[5] Whether addressed to Old Norse deities or to the Christian God, the formula implies an emic understanding that divine or magical forces could be invoked to influence seasonal rhythms and environmental balance. Its presence in both religious systems illustrates how traditional ritual language was absorbed rather than erased in post-conversion Scandinavia, which key feature of what Mitchell (2011: 44) refers to as cultural loan shift.

Although many medieval theologians condemned non-Christian religion and practices, many non-Christian elements, like weather magic, were adopted and transformed by Christians into acceptable means of religious expression. Flint (2002: 118) suggests that in early medieval Europe, 'counter-magi' figures were developed by the Church, who performed duties that were similar to the local non-Christian traditions of magic. She further puts forward that the replacement of magic with miracles was a conscious effort by local religious figures, as the majority of lay-people still depended on non-Christian magicians. This deliberate replacement of magic performers with 'counter-magi' Christians is certainly plausible in medieval Scandinavia, where bishops and saints are portrayed in medieval texts as being able to perform miracles that echo the nature-bending magic of non-Christian magic, like weather magic. After the official conversion to Christianity, people in medieval Scandinavia still relied on the services that magic provided enough to seek magic performers, which is demonstrated in the twelfth-century law code, *Eiðsivaþingslǫg*. As demonstrated earlier in the previous chapter on Indigenious weather magic (Chapter 4), in this legal text, residents in south-eastern Norway were forbidden from travelling to Finnmǫrk and to seek Saami magicians. The need to legally ban people from travelling to seek out magic solutions suggests that in twelfth-century Norway, there were still a significant number of people who relied on magical experts. Flint suggests that:

> [i]f the Christian church were to have any lasting impact on such persons[...] then it must stop just condemning them and their magicians wholesale, and attend instead to the needs that the magicians fulfilled, this with a view to offering a viable alternative.
>
> *(Flint 2002: 118)*

Flint (2002: 130) further argues that the medieval Church dealt with non-Christian magic by seeking to outdo it through the absorption of these acts of non-Christian magic. This integration of non-Christian magic into Christian practices is apparent in post-conversion medieval Scandinavia. Raudvere (2002: 91) posits that the non-Christian emphasis on performing magical acts through the spoken word is one example of cultural continuity that occurs throughout non-Christian Scandinavian society, and that some of these verbal acts were adopted into a Christian tradition. Often, non-Christian weather-magic performances conducted through speech were transformed into Christian acts that were conducted through spoken prayers, blessings, and vows. This absorption of non-Christian magic can perhaps be seen in Adam of Bremen's portrayal of Saint Rimbert, where the Christian saint is depicted as having the ability to still stormy seas through spoken prayer during his travels to Sweden (Adam of Bremen 2002: 40).[6] It is necessary to acknowledge that for a largely oral society like the one in medieval Scandinavia, it is unsurprising that there would be a reliance on the spoken word in any ritual. However, the example of Saint Rimbert highlights a cultural emphasis on the spoken word in combination with the outcome of weather change, and that emphasis is what is really important to the Scandinavian tradition of weather magic and miracle. In the example of Saint Rimbert, alongside the majority of examples of weather miracles discussed in this chapter, the emphasis is placed on the vocal performance resulting in a change in the weather, rather than including any of the physical rituals that are typical of medieval Catholicism. The possible existence of a cultural continuity, like this emphasis on the spoken word with an outcome of a change in weather, is not to say that Christianity did not already have traditions of weather miracles in regions outside Scandinavia,[7] but that aspects of non-Christian Norse weather magic were adopted into the medieval Scandinavian interpretation of Christianity. This shift in belief occurs through a form of religious syncretism,[8] or the blending of two religious systems into a new belief system, which Stephen Mitchell refers to as a 'cultural loan shift' (Mitchell 2011: 44). Rather than simply representing the continuation of non-Christian practice following Scandinavia's conversion to Christianity, Mitchell suggests that a cultural loan shift, in reference to medieval Norse magic specifically, is the adaptation of non-Christian customs and traditions into the new, Christian interpretive framework.

When considering weather magic, the likelihood of a cultural loan shift is supported in that the daily reliance on weather transcends Scandinavia's change from a predominantly non-Christian society to a predominantly Christian society. Scandinavian sailing and agricultural communities would have remained heavily reliant on the weather for their livelihood, regardless of the principal religious belief. This remainder of this chapter will analyse portrayals of non-Christian weather magic found in Old Norse saga literature, alongside the depiction of weather miracles in similar Old Norse textual sources, to highlight potential cultural loan shifts.

The medieval Scandinavian reliance on favourable weather for the success of their agricultural efforts is demonstrated by the presence of weather miracles found throughout Old Norse and continental texts. In *Gesta Hammaburgensis ecclesiae pontificum*, Adam of Bremen describes a certain Bishop Adalward the Elder as a 'truly praiseworthy man' (Adam of Bremen 2002: 205), who preaches to the non-Christian communities of Sweden whilst living in Götaland. Adam notes that Bishop Adalward was especially known for his miraculous powers, including his ability to manipulate the weather. One of the bishop's weather miracles is demonstrated in Adam's text when a non-Christian community asks for rain and Bishop Adalward brings a rainfall and then returns fair weather to the area. Bishop Adalward's noted reputation for weather miracles highlights the importance that medieval Scandinavian communities placed on religion to ensure agricultural success, and in turn, safeguarding their livelihood.

In a similar episode from the fifteenth chapter of *Jóns saga ins helga*, an early thirteenth-century saga recounting the life of Bishop Jón Ǫgmundarson (1052–1121), Jón performs a weather miracle that brings an end to a prolonged winter season followed by a summer of drought. After a particularly dry spring, Jón arrives at the *alþingi* in June and promises an end to the dry spell. After his promise is made, rain begins to pour and the harvest is saved:

> Sumarit hafði verit þurrkusamt áðr svá at eigi hafði getit dǫggvar, ok váru fyrir því grǫsin nær engi. Ok þá er inn helgi Jón byskup hét, þá ver svá þunnt veðr ok hreint at hvergi sá ský á himni. En þá er hann hóf upp Gloria in excelsis þann dag í messu, þá kom regn svá mikit af himni at

mönnum þótti illvært úti fyrir vátviðris sǫkum. Þaðan ífrá um sumarit váru dǫggvar um nætr en sólskin um daga.

(ÍF 15: 230)

The summer had been dry until then and there has been no dew, and for that reason there was hardly any grass. And when the holy bishop John called on God for help, it was such dry, clear weather that not a cloud was to be seen in the sky. And when he sang the Gloria at mass that day, then so much rain fell from the heavens that people thought it an ill thing to be outside because of the wet weather. From then on that summer, there was dew at night and sunshine during the day.

(McCreesh 2018: 26)

As scholars have previously pointed out (McCreesh 2018: 26; Langeslag 2015: 24), many of the weather miracles concerned with agricultural success that are represented in the *biskupasǫgur* are not unique to Old Norse texts. In particular, McCreesh (2018: 26) questions the originality of Jón's weather miracle from *Jóns saga ins helga* because it follows the weather-miracle motif from Gregory the Great's *Dialogues*:

The sky was so clear at the time that there was not a cloud in sight. At her brother's refusal Scholastica folded her hands on the table and rested her head upon them in earnest prayer. When she looked up again, there was a sudden burst of lightning and thunder, accompanied by such a downpour that Benedict and his companions were unable to set foot outside the door.

(Gregory the Great 1959: 102–103)

McCreesh (2018: 26) and the editors of the ÍF edition of *Jóns saga ins helga* (ÍF 15: 320) comment on the similarities of a clear sky and sudden onset of rain that is so strong that people refuse to go outside, which is portrayed in both Scholastica's and Jón's weather miracles. Régis Boyer (1973: 152) notes that Gregory's *Dialogues* were popular in medieval Iceland and could have inspired many of the compilers of Old Norse saga literature, so it is likely that the compiler of *Jóns saga ins helga* would be familiar with Scholastica's weather miracle. Regardless of whether Jón's weather miracle scene was influenced by Gregory's *Dialogues*, there are certain aspects of Jón's weather manipulation that are uniquely Icelandic. Firstly, the motivation for the miraculous rain in *Jóns saga ins helga* is completely different from the motivation for the change in weather in Gregory's *Dialogues*. In *Dialogues*, Saint Scholastica and her brother, Saint Benedict, met once a year to spend a day together discussing spiritual matters. On the last day that they met before Scholastica's death, she begged her brother to stay through the night and continue to speak with her about God. Benedict refuses, reminding

Scholastica of the monastic rule that he would not sleep outside of the cloister. It is then that Scholastica prays to God and the torrential rain comes, stopping Benedict from leaving the abbey. As demonstrated in the above quote from *Jóns saga ins helga*, Jón is reacting to a drought, which is a common motif that is found in medieval Icelandic literature. Langeslag (2015: 24) emphasises that what makes the weather miracles of the *biskupasǫgur* as a whole particularly Icelandic is how they collectively highlight the precarity of Iceland's grazing seasons, and how devastating a prolonged winter season could be on communities' livestock. This is certainly the case for Jón's weather miracle, as the motivation for the miracle stems from a lack of rain that was causing the grazing grass to die. I would also suggest that, although it is possible that Jón's weather miracle could have been inspired by the weather miracle found in Gregory's *Dialogues*, the description of how Jón performed the weather miracle is more unique to Scandinavia. In the weather miracle that is depicted in *Dialogues*, Scholastica brings on the rain by folding her hands, resting her head and praying, without any specific mention of whether she prays out loud. Nothing about Scholastica's act of prayer is connected to Scandinavia, whereas Jón's performance of his weather miracle places the emphasis on his vocal performance resulting in the rain, which, as I have shown, is common in Scandinavian weather-magic tradition. In a markedly different style of performance, Jón calls on God for help and sings the Gloria during mass, and both acts place an emphasis on the bishop's vocal performance. As demonstrated in the previous chapter on weather-magic performances, the spoken word played an important role in the performance of non-Christian weather magic. The emphasis on the vocal aspect of Jón's weather miracle performance could suggest a cultural loan shift from the non-Christian emphasis on the spoken word in weather-magic performances.

A second possible loan shift is perhaps demonstrated through the importance of controlling the weather while at sea, which is demonstrated in both non-Christian weather magic and Christian weather miracles. In *Þorláks saga byskups*, it is said that whenever people were faced with perils resulting from the weather, individuals called upon Bishop Þorlákr and he could still a storm, subside the winds and the calm the sea (Þorláks: 32). Þorlákr's reputation for helping sailors in dangerous weather is emphasised in the forty-eighth chapter of the same saga, when a crew sailing out from the Vestmannaeyjar are caught in a violent storm and call to the bishop for help: 'Many went by ship out from the Vestmannaeyjar and a rough sea and a storm came upon them so that they lay near death. They then called upon Bishop Þorlákr to help them and at once a great calm fell and the sea became tranquil and they reached harbour safely' (Þorláks 2013: 27). Þorlákr's usefulness in tumultuous weather was not limited to those sailing in Icelandic waters. As Chapters 63 and 64 of *Þorláks saga byskups* reveal, Norse sailors could also call upon Bishop Þorlákr when they found themselves in storms as far away as the Faroe Islands and the English Channel.[9]

However, the ability to miraculously calm storms at sea is not unique to Þorlákr, nor is it unique to Old Norse texts. As referenced above, Óðinn is also represented as a religiously significant figure who could calm the seas when sailors were in need. Yet, the lack of description surrounding either Þorlákr or Óðinn's performances when manipulating the weather on behalf of sailors makes it impossible to say for certain whether Þorlakr's ability to calm sea storms derives from a non-Christian influence. Furthermore, there are other examples of weather miracles that calm storms at sea occurring outside medieval Scandinavia. Perhaps the most well-known examples of sailing miracles that do not stem from a Scandinavian tradition are those attributed to Jesus, as portrayed in the Gospels.[10] For example, the Gospel of Mark (4:35–41) includes an episode where Jesus and his disciples cross the Sea of Galilee, when the boat is engulfed in a violent storm. Jesus remains asleep as the waves crash into the boat and water begins to build up in the shift, so the disciples ask Jesus if he is going to let them drown. Jesus wakes up and calls for the winds to be still. The winds immediately cease, and the sea becomes still. Therefore, it is possible that Saint Þorlákr's weather miracles at sea stem from a non-Scandinavian tradition.

The many examples of weather miracles found in Old Norse texts suggest that, as Flint proposes (2002: 118), it is plausible that the local church authorities in medieval Scandinavia consciously replaced the reliance on non-Christian weather magic with 'counter-magi' figures, who were also able to manipulate the weather through the power of God. The portrayals of miraculous weather changes in the primary sources demonstrates a continuity between how acts of non-Christian weather magic and Christian miracles were thought to be performed. These potential cultural loan shifts, like the reliance on vocal performance, are what make the weather miracles found in Old Norse saga literature uniquely Scandinavian, in addition to the Norse-centric motivations for such weather miracles, such as famine and safety at sea.

Conclusions

This chapter has demonstrated the nuanced distinctions and overlaps between weather miracles and weather magic as they are depicted in medieval Scandinavian texts. Through an examination of episodes of weather miracle and weather-magic performances found in both continental and Old Norse sources, several key conclusions emerge about the nature and representation of these practices during the medieval period.

One of the primary insights is the intricate distinction between weather magic and weather miracles. Contrary to the binary opposition suggested by some scholars (i.e. McCreesh 2018, 2006; Meylan 2011), the boundaries between these practices are not rigid but rather complex and fluid. For example, in *Ólafs saga Tryggvasonar*, Bishop Sigurðr employs a combination of prayer, holy water, and religious symbols to counteract Rauðr inn rammi's weather magic, enabling King Ólafr's ships to navigate through a fjord. This episode highlights the detailed and ritualistic nature of Sigurðr's actions, which closely resemble non-Christian magical practices. The meticulous documentation of each step taken by the bishop to overcome the magical storm underscores a sophisticated understanding and the importance of ritual within a medieval Christian Scandinavian context.

Similarly, the portrayal of Guðmundr Arason in BS II reflects this complexity. His prayers for favourable winds during a sea voyage are depicted with a focus on his physical and vocal actions, echoing the performative aspects of non-Christian weather magic. These examples challenge the simplistic categorisation of weather miracles as purely divine and inherently righteous, demonstrating instead a blend of ritualistic practices that blur the lines between magic and miracle.

The chapter also provides substantial evidence supporting the argument that Christian weather miracles were influenced by non-Christian weather magic. This cultural loan shift is particularly evident in the emphasis on vocal performance. In *Jóns saga ins helga*, Bishop Jón Ǫgmundarson twice uses prayer to alter the weather, bringing an end to both a drought and a prolonged winter. The focus on vocal elements, such as Jón singing the *Gloria* during mass, mirrors the performative nature of non-Christian weather magic, where spoken words were central to the performance.

Moreover, the episode from *Ketils saga hængs* demonstrates a continuity in the performative aspects of weather manipulation. The local Indigenous community that features in this scene uses spoken incantations to summon a storm, protecting their livelihood from external threats. The vocal performance is a critical component of the magical performance, underscoring the importance of spoken words that is repeatedly emphasised in both non-Christian and Christian literary contexts. This continuity suggests that the Christianisation of Scandinavia did not entirely erase non-Christian practices but rather integrated and reinterpreted them within a new religious framework.

Another significant finding is the distinction between the intended beneficiaries of weather miracles and weather magic. Weather miracles, as depicted in the *biskupasǫgur*, were often performed with an aim to benefit the wider community. For instance, in *Jóns saga ins helga*, Bishop Jón Ǫgmundarson's prayers for rain save an entire community from drought, ensuring the success of their agricultural efforts. Similarly, in *Guðmundar saga biskups*, Guðmundr Arason's prayers for favourable winds benefit not only his own safety but also that of his companions and the broader community relying on safe sea travel.

In contrast, weather magic in Old Norse saga literature frequently serves individual purposes. In *Gísla saga Súrssonar*, Auðbjǫrg's summoning of a snowstorm to cause an avalanche is motivated by personal vengeance against her neighbour, whose actions had injured her son. This act of weather magic benefits only Auðbjǫrg (and perhaps her son), illustrating the personal and often selfish motivations behind such practices. However, there are exceptions to this general trend. In *Vatnsdœla saga*, Bárðr stirfinn uses weather magic to calm a blizzard, benefiting his entire community. It is noteworthy that Bárðr and his community are described as Christians, suggesting a possible syncretism where Christian identity does not preclude the use of weather magic for communal benefit.

The Old Norse sagas frequently portray Christian figures as possessing the power to overcome non-Christian magic, reinforcing the narrative of Christian superiority in the post-conversion, medieval Scandinavian worldview. This is evident in the episode from *Ólafs saga Tryggvasonar*, where Bishop Sigurðr's Christian rituals successfully counteract Rauðr's weather magic, and calm the stormy weather. The detailed description of Sigurðr's actions serves to highlight the efficacy and superiority of Christian practices. Similarly, in *Þorláks saga byskups*, Bishop Þorlákr is repeatedly depicted as calming storms at sea, a power traditionally associated with non-Christian Old Norse deities like Óðinn. Þorlákr's ability to calm the seas not only aids individual sailors but also reinforces the narrative of Christian dominance over nature and non-Christian magic. These portrayals serve a dual purpose: they validate the Christian faith by demonstrating its superior power and provide a framework for integrating non-Christian practices into a Christian worldview.

The specific motivations and circumstances surrounding weather manipulation practices reflect the particular needs and challenges of medieval Scandinavian society. The reliance on favourable weather for agricultural success and safe sea travel is a recurring theme in both weather magic and weather miracles. In *Jóns saga ins helga*, the motivation for Bishop Jón's weather miracles is rooted in the need to save crops from drought, reflecting the agricultural concerns of medieval Icelandic communities.

Similarly, the emphasis on safe sea travel in *Þorláks saga byskups* underscores the importance of maritime activities in Scandinavian society. Bishop Þorlákr's ability to calm storms benefits sailors navigating dangerous waters, illustrating the practical concerns that motivated weather miracles. These context-specific adaptations highlight the pragmatic approach to weather manipulation in medieval Scandinavia, where the primary concern was often the immediate needs of the community or individual, regardless of the religious or magical classification of the practice.

In conclusion, the examination of weather miracles and weather magic in medieval Scandinavian texts reveals a complex and nuanced cultural understanding where the boundaries between magic and miracle are fluid and often overlapping. The integration of non-Christian elements into Christian practices, the differentiation between community and individual benefits, and the portrayal of Christian superiority all contribute to a multifaceted understanding of weather manipulation in this period. The continuity and change in performances, along with context-specific adaptations, further underscore the pragmatic nature of these practices. By recognising these complexities, we gain a deeper insight into the cultural and religious dynamics of medieval Scandinavia and the ways in which communities navigated the challenges of their environment through both magic and miracle.

Notes

1 McCreesh (2018: 24) highlights that Walter Gerhardt Jr (1966) even goes so far as to argue that God as represented in the Old Testament was originally a weather god.

2 See Chapter 1 on weather-magic terminology.

3 See Chapters 1 and 3 of the present study for previous discussions surrounding this scene in *Vǫlsunga saga*.

4 See Chapter 2 for more discussion on this scene from *Gísla saga Súrssonar*.

5 The phrase *til árs ok friðar* appears in multiple contexts, including ritual formulas, legal texts, and Old Norse saga literature. For example, it is included in the *Gulaþingslǫg* and *Grágás* law codes as part of sacrificial or legal phrasing (see *Grágás* Ia: 15, 148), and is also preserved in literary sources such as *Hákonar saga góða*, where it is associated with pre-Christian ritual sacrifice to secure prosperity and peace (ÍF 26: 126–127). Its continued use into the Christian period suggests a degree of cultural continuity across religious transitions.

6 Interestingly, Saint Rimbert himself recounts another episode of a weather miracle from his missionary trip to Scandinavia, in his biography about Ansgar, the *Vita Ansgari* (Waitz 1884). In this story, a non-Christian community is drenched by a storm, while the Christian, Herigar, remains dry. According to Rimbert, Herigar uses this as a demonstration for his argument that the Christian god was more powerful than the non-Christian gods (Waitz 1884: 40).

7 For examples of Christian weather magic in Classical Antiquity, see McCartney (1925).

8 For an overview of the various forms of syncretism, see Mitchell 2011 and Berner 2001.

9 Chapter 63: 'Some merchants from Orkney were driven off course by a storm to the Faroes where there were cliffs and reefs, and it seemed certain death to all of them. They called on the holy Bishop Þorlákr and at once the storm turned into a favourable wind around them' (Þorláks: 28).

Chapter 64: 'Other merchants had come to the point of death in the English Channel because of the storm and the height of the sea. They called upon Bishop Þorlákr and at once got a fair wind and balmy weather and there-after a good harbour' (Þorláks: 28).

10 The miracles of Jesus calming storms are depicted in Matthew 8:23–27, Mark 4:35–41, and Luke 8:22–25.

6

WEATHER MAGIC AND ELITE POWER

Representations of Óðinn in Medieval Scandinavia

This chapter examines how Óðinn is represented in medieval Scandinavian textual sources as a figure skilled in weather magic, with particular focus on depictions of his ability to influence winds and storms at sea. The discussion is intentionally cultural-historical rather than theological in scope: it does not seek to define Óðinn as 'the' weather god, nor to position him in opposition to Þórr or other storm-associated figures, but instead explores how weather magic associated with Óðinn functioned within elite discourse, poetic imagery, and medieval understandings of magic and power. Although scholars have noted aspects of Óðinn's association with wind and weather (e.g., Perkins 2001: 16–17; McCreesh 2018: 57–58; Price 2019: 57), much of the academic discussion of Norse deities and the weather has concentrated on Þórr (e.g. Perkins 2001; Bertell 2003; Taggart 2017). What remains missing is a study that examines the attribution of weather manipulation to Óðinn in depth, bringing together the full range of literary and poetic evidence that connects him with weather magic.

Perhaps stemming from Adam of Bremen's and Snorri's taxonomy of the Norse gods, earlier scholarship is often limited to an analysis of the gods' roles within set functions. Georges Dumézil (1973) organised the Old Norse religion into a 'trifunctional' system, which creates a hierarchical society consisting of the higher class (rulers and priests), warriors, and common people, and then categorises religious functions into sovereignty, military, and productivity (e.g., farming and fertility). Dumézil's taxonomy heavily influenced Old Norse religious scholarship produced throughout the second half of the twentieth century, which resulted in scholars adhering to the 'set functions' originally given to certain Old Norse gods. For example, when glossing Óðinn's name 'Viðrir', Rudolf Simek (1993: 360) suggests that although Óðinn often appears repeatedly under this name associated with the weather,

DOI: 10.4324/9781003746058-9

he is not 'the actual weather god of Nordic mythology', even though he occasionally influences the weather through magic. Simek is, of course, referencing Þórr as 'the' weather god. However, within the past decade, the field of Religious Studies has begun to deconstruct the notion that the Norse gods formed a fixed pantheon, while also acknowledging a fluidity in the roles that each god played in pre-Christian society, depending on who was doing the worshipping, and where. As Terry Gunnell puts it:

> Old Norse religious beliefs and practices varied by time and space, largely in accordance with social practices and environments, depending on whether people lived in the mountains, or in the flat agricultural lands, by the sea, or on islands; depending on whether they were farmers, warriors or hunters; and also depending on the degree of social contact they had with people outside of the immediate Nordic area.
>
> (*Gunnell 2015: 57*)

This fluidity in functions, shaped by the worshipping community in question, is particularly relevant when considering how weather manipulation is attributed to the Norse gods in the sources. At the same time, this chapter deliberately moves away from functional models altogether, approaching weather magic not as a stable divine role but as a culturally produced attribution preserved in medieval literary and poetic traditions. It therefore focuses on Óðinn as a figure to whom weather magic is attributed in the surviving sources.

This chapter will concentrate on Óðinn as he is represented in the extant medieval textual sources, such as Old Norse saga literature, eddic poetry, and skaldic poetry, which mainly originate from Iceland. Most historians of religion recognise that the Óðinn presented in these surviving sources is what Schjødt (2019: 59) describes as an 'end result' of the god, which has gone through centuries of development and change.[1] Indeed, many scholars of religious history generally agree that around the year 500 AD, extensive changes in beliefs and approaches to religious practice began to take place in southern Scandinavia (see Gräslund 2007 and Andrén 2014). However, it would be impossible to attempt to reconstruct a complete history of Óðinn as a weather-manipulating god in the present study. Instead, the discussion that follows examines how weather manipulation is attributed to Óðinn in eddic and skaldic poetry, supported by Old Norse saga evidence and relevant continental parallels, in order to explore how divine weather magic was imagined as part of elite discourse and poetic expression.

Óðinn, a Multi-Faceted God

As Jens Peter Schjødt (2019: 59) explains, Óðinn is a multi-faceted and complex figure within the medieval sources. Scholars focusing on Norse religion and mythology include war, poetry, seduction, and magic practitioner among the many attributes that Óðinn is portrayed as having in the Viking and Middle Ages (Lassen 2021; Price 2019, 2002; Gunnell 2017, 2015; Kershaw 2000; Simek 1993; De Vries 1970). These traits are not set functions in that the god would have been limited to fulfilling only these roles, nor do all of the attributes consistently appear in every source that refers to Óðinn. Furthermore, Óðinn is not the only Norse god to have been linked to depictions of weather manipulation in the sources. Many other divine and semi-divine figures, including Þórr, Þorgerðr Hölgabrúðr, and Irpa, also appear as storm-controllers in medieval texts. Rather than isolating one 'storm god' or 'weather god', this chapter considers Óðinn within this broader constellation of beings associated with weather magic. There is a wealth of medieval source material that connects the worship of Óðinn to the elite and warrior groups of society, yet he is most often represented in the textual sources as a companionless wanderer (Price 2019: 55; Schjødt 2008: 220). Neil Price (2019: 57) emphasises that the realm of magic is present in virtually all aspects of Óðinn, including his role on the battlefield, and that Óðinn was a god for both the old and young, balancing his reputation for aggressiveness and of wisdom. In what follows, I focus on the imagery that links Óðinn to storms and weather magic. Much of this appears in the figurative language and layered meanings of skaldic kennings, which do not indicate a theological category of storm deities, but rather show how representations of divine weather magic were mobilised as a language of elite power and authority.

It is fairly safe to say that the majority of scholars agree that Óðinn was particularly relevant for the elite and warrior groups of Scandinavia and played an important role in battle and warfare. In *Gesta Hammaburgensis ecclesiae Pontificum* (c. 1073–1076), Adam of Bremen offers a description of the temple in Uppsala, inside which he says sit statues of three gods: Þórr, Wodan (Óðinn) and Fricco (Freyr). Adam presents Óðinn as 'the Furious' god and further describes how he 'carries on war and imparts man strength against his enemies' (Adam of Bremen 2002: 207). Adam's narrative then provides a portrayal of how each of the gods are supposedly depicted in their statues, with Óðinn in armour (which, as Adam notes, is similar to how his contemporaries depicted Mars in armour). Adam continues with an explanation of the appointed priests who were assigned to the temple at Uppsala, and their role in offering sacrifices to the three gods in particular scenarios. It is noted that if a war was taking place, a sacrifice should then be made to Óðinn (Adam of Bremen 2002: 207). However, Adam had never been to Sweden himself and relied on second-hand accounts to formulate his depiction of the temple in Uppsala. Fitting three gods into three set functions is also problematic, as Týr, who is more often seen as a counterpart of Mars in the tripartite system, is no longer included (Belier 1991: 103).

A euhemerised version of Óðinn is presented in a collection of tales incorporated in *Ynglinga saga* (c. 1220), which echoes Adam's portrayal of Óðinn as a war-focused god. In the sixth chapter of *Gylfaginning*, Óðinn is described as the highest and most ancient of all the Æsir. Alongside eleven other names, he is referred to as Valfǫðr, or father of the slain, and is intrinsically connected to warriors in battle. However, it is crucial to keep in mind that the presentation of the Old Norse religion and its gods comes in these texts from a Christian perspective, reflecting on a pre-Christian religion over two hundred years after Iceland's official conversion to Christianity. This extensive gap in time, alongside the texts' Christian worldview, leaves room for aspects of the pre-Christian religion to have either been altered or forgotten. The Óðinn presented in the *Prose Edda* is by no means an exact, comprehensive image of what Óðinn would have been worshipped as in the Viking Age.

As can be seen from the above, there are no obvious indications in either *Gesta Hammaburgensis ecclesiae Pontificum* or *Gylfaginning* that Óðinn was conceptualised or invoked primarily in relation to weather within cultic practice. All the same, it is not surprising that skaldic poets present Óðinn through imagery associated with warfare and stormy, chaotic weather. Both harsh weather and warfare invoke a sense of chaos and are often portrayed as inherently connected throughout the Old Norse corpus. Bernadine McCreesh (2018: 137) notes that the weather was often portrayed in the *Íslendingasǫgur* as a malignant force that, like any opposing force, could be attacked and defeated. In short, fighting in a battle or surviving a storm at sea can be perceived as similar in that once fully enveloped in the chaos of the event, the outcome of either scenario could seem out of one's own control without the influence of a divine power or perhaps magical intervention. This connection between frenzied battles and tumultuous weather is often perpetuated in Old Norse eddic poetry, skaldic poetry, and saga literature. This idea has been underlined by Hannah Burrows (2021) who also considers the parallel between harsh weather imagery and scenes of warfare by suggesting that just as winter weather brings sterility to lands, warfare can cause similar ruin and destruction. Such a chaotic situation might fit well into those spheres connected with Óðinn.

Óðinn as a Weather-Magic Practitioner

Depictions of Óðinn controlling the weather are particularly relevant to this study, as the weather manipulation directly attributes Óðinn's ability to change the weather to his knowledge of magical chants. Óðinn is certainly presented in the surviving textual sources as a central figure through whom knowledge of magical practice is repeatedly articulated, a connection which is highlighted in the extant Old Norse poetry and prose.

The god's relationship with magic is perhaps most clearly exhibited in the eddic poem *Hávamál*. The final section of this poem, which includes a list of metrical spells that Óðinn knows, is named in modern scholarship *Ljóðatal* or 'list of chants'. Carolyne Larrington (1993: 62) suggests that *Ljóðatal* is an index of spells used to fulfil the literary functions of illustrating the different types of wisdom and unifying the different elements that make up the character of Óðinn. The eighteen spells listed in *Hávamál* provide Óðinn with aid in a variety of scenarios, such as curing illnesses, communicating with the dead, providing bravery on the battlefield, and controlling the weather. The god's knowledge of magic is echoed in other surviving works of medieval prose, such as in *Ynglinga saga* (Ch. 7), in which Óðinn is said to have knowledge of magic runes, *galdrar*, and *seiðr*, the last of which allows him to predict the fates of men and cause death and disease.

As Perkins (2001: 16) briefly notes in his study of Þórr as a wind god, Óðinn's ability to magically control the elements is demonstrated in many sources. Perhaps the most well-known attestation of this is found in the aforementioned *Ljóðatal*, in which the god is presented as knowing a *galdr*, or magical chant, that will quieten the winds at sea:

154 Þat kann ec iþ
ellipta, ef mic
nauðr um stendr,
at biarga fari míno á floti:
vind ec kyrri vági á
oc svæfic allan sæ
(Neckel & Kuhn 1962: 43)

I know an [eleventh] one if I am
in need, if I must save my ship
when afloat;
the wind I can quieten upon the wave
and lull all the sea to sleep
(Larrington 2014: 34)

This eleventh spell is presented as a means of weather magic that Óðinn knows, and uses, to save ships caught in a storm. This weather-magic knowledge is echoed in *Ynglinga saga* (Ch. 7), where Óðinn is also depicted as having magical control over the elements: 'Þat kunni hann enn at gera með orðum einum at sløkka eld ok kyrra sjá ok snúa vindum hverja leið, er hann vildi' (ÍF 28: 18) ['[Oðinn] also knew how to put out fire or calm the sea or turn the winds in any direction he wished with words alone' (*Heimskringla* I: 10)]. In both of these sources, Óðinn's knowledge in weather magic appears particularly relevant to seafaring. This link between Óðinn's weather-magic performances and sailing is perpetuated in many Old Norse sources, as for example, in the Eddic poem *Reginsmál*. In this poem, Sigurðr and his men encounter a storm at sea when sailing to meet the sons of Hunding in battle. Fleeing from the storm, Sigurðr and his crew sail along the shoreline and then meet Óðinn, disguised under the names Hnikarr, Feng, and Fjǫlnir, who comes on board their ship and calms the storm:

18 Hnicar
héto mic,
þá er hugin
gladdi
Volsungr ungi oc
vegit hafði Nú máttu
kalla karl af bergi,
Feng eða Fiǫlni far vil ec þiggia.

Þeir vico at landi, oc gecc karl á skip, oc lægði þa veðrit.

(Neckel & Kuhn 1962: 178)

Hnikar they called me, when young
Volsung gladdened the raven when
there was fighting; now you may call
the old man on the cliff Feng or Fiolnir;
I would like passage!

They turned towards the land, and the old man went aboard the ship, and the tempest abated.

(Larrington 2014: 151)

This scene is also represented in *Norna-Gests þáttr*[2] (Ch. 6) and *Vǫlsunga saga*[3] (Ch. 17). In the three sources that include this episode, the origin of the storm and its potential connection to magic are disputed. The narrator's inclusion of *gerningaveðr*, a compound term that attributes bad weather to magic that was discussed in the first chapter of the present study, in *Norna-Gests þáttr* certainly indicates that the tempest is the product of magic: 'Þa fíngv ver gerninga veðr stor. Þat kendv menn Hvundings sonvm' (Ólafur Halldórsson 2000: 24) [Then we got a large storm raised by magic, and people knew that it was the sons of Hunding (own translation)]. In both *Reginsmál* and *Vǫlsunga saga*, however, the actual origin of the storm is not attributed to weather magic. In the former, the storm is portrayed as a '*storm mikinn*' (Neckel & Kuhn 1962: 177) [great storm], with no magic implied. In *Vǫlsunga saga*, the abrupt turn in weather is narrated as follows: '*þá kom á veðr mikit með stormi*' (Finch 1965: 28) [a violent storm of wind came up (Finch 1965: 28)].

Regardless of whether the storm's origins were magical or not, in each of the portrayals of this scene, Óðinn is the one responsible for calming the storm. In *Reginsmál*, the prose following the eighteenth stanza reveals that once Sigurðr welcomes Óðinn onto the ship, the violent storm abates. Although the poem does not directly state that the change in weather is a result of Óðinn's knowledge of weather magic, there is the implication that the weather turns favourably as a result of Óðinn being welcomed onto the ship. The change in weather is more firmly tied to Óðinn's arrival in *Norna-Gests þáttr*'s representation of the episode: 'Þa vikv ver at landi, ok lægdi skíott vedrít ok bad Sigurðr karl ganga vt á skíp. Hann gerdi sva. Þá fell þegar vedrit, ok gerði ín bezti byr' (Ólafur Halldórsson 2000: 25). [Then they turned ashore and the weather immediately calmed, and Sigurðr asked the man to come out to the ship. He did so. Then the weather fell at once and made a favourable wind (own translation).] Here, the violent weather dies down directly after Sigurðr welcomes Óðinn onto the ship, and then is turned into favourable wind. The connection between Óðinn and the change in weather is also firmly made in *Vǫlsunga saga*,[4] where the storm abates as soon as Óðinn is welcomed on board, and the crew are then able to sail smoothly to the kingdom of the sons of Hunding. Once the ship arrives to its destination safely, Fjǫlnir disappears.

In the late twelfth-century *Gesta Danorum*, the Danish historian Saxo Grammaticus also provides an indirect connection between Óðinn and his ability to magically manipulate the weather at sea. In the work's account of Starkaðr's life (Saxo vi.5.7), an expedition of raiders finds themselves stuck on land for almost a year, due to adverse sailing conditions. The narrative implies that the lack of favourable winds is a result of Óðinn's plan for Starkaðr to bring about the death of Starkaðr's sailing companion, Víkarr, who was also king of Norway. After several months pass with no change in the weather, Starkaðr reveals that the only way to continue on their journey is to offer a human sacrifice to Óðinn, in the hope that the god will then grant them better sailing conditions. All of the men agree and cast lots into an urn, which results in King Víkarr being selected for the sacrifice. Starkaðr tricks Víkarr into an early grave by convincing the king that his sacrifice will only be a symbolic offering using a noose made of willow, rather than using a stronger material that could actually kill Víkarr. However, Starkaðr purposefully ties the willow noose so tight that Víkarr dies, and the sacrificial act is complete. Following Víkarr's death, the winds change, and the fleet is able to continue on their journey. Although Saxo Grammaticus' narrative never explicitly attributes the change in weather to Óðinn's divine intervention, the sailors are portrayed as believing that Óðinn is capable of providing a change in winds. Regardless of whether the episode reflects Víkarr's death with any accuracy, or even if Saxo himself was unconvinced of the historicity of sacrificing to Óðinn for favourable winds, the narrative perpetuates the cultural connection between Óðinn and weather magic.

This scene is also portrayed in the later medieval *Gautreks saga* (Ch. 7), where Víkarr, Starkaðr and the rest of the raiding party are stuck in Hǫrðaland due to adverse winds. Similar to the episode in *Gesta Danorum*, the crew predicts that Óðinn will demand a sacrifice in order for the weather to change. In this legendary saga, Starkaðr meets Óðinn and Þórr in person later that night, and it is here that Óðinn clearly demands Víkarr's sacrifice from Starkaðr. Yet rather than the sacrifice being made to change the weather, in this episode, Óðinn's demand for a sacrifice is a means of sealing the blessings and curses that the gods have placed on Starkarðr.[5] Once again, through deception on Starkarðr's end, Víkarr is sacrificed the next day and in this version of the story, Starkarðr is then outlawed from Norway.[6] Although Víkarr's sacrifice is not explicitly made to change the weather, the sailors' initial prediction that Óðinn must have required a sacrifice is made as a reaction to the adverse sailing conditions. This episode in *Gautreks saga* once again perpetuates a notion that Óðinn was culturally associated with having control over sailing weather.

Óðinn's Control Over the Weather in Skaldic Poetry

The mythological examples discussed above that reference Óðinn's ability to control the weather only exist in manuscripts compiled after the Viking Age, in which the political and religious agendas of the authors and scribes must be considered. Although Old Norse literature often roots in oral culture, the gap in time and conversion to Christianity has inevitably resulted in alterations to the mythology, and there is no fool-proof way of checking what information recorded in these textual sources has gone through changes because of time passing and a change of religious beliefs. Eddic poetry is similarly problematic in that without the strict metrical structures that exist in skaldic compositions, these Eddic poems may also have been subject to alterations over time. Unlike Eddic poetry and the Old Norse saga literature, it has been argued that much skaldic poetry can be more securely traced to the Viking Age. Whereas the Eddic poems are anonymous and normally focus on 'timeless' subject matter, skaldic verse is mostly attributed to specific poets and composed for a particular patron, often as a means of commemorating a particular event (Ross 2005: 69–82). Knowing this information helps scholars to date the poems with more accuracy, and the metrical structures ensure fewer alterations in the verse content (Jesch 2008: 295).

The majority of skaldic references to Óðinn's control over the weather are preserved in the intricate circumlocutions known as kennings. These kennings are normally constructed from two or more words that represent a single noun. As an example, the phrase 'candle of the sky' would be a kenning for the sun. The kennings that are formed using a connection between Óðinn and weather normally fit into particular kenning patterns that indicate warfare. An example is found in *Hákonarmál*, a poem that is attributed to Eyvindr skáldaspillir Finnsson. In the eighth stanza, the kenning *Óðins veðr*, or 'the storm of Óðinn' is used:

8. Blendusk við
roðnar und
randar himni;
Skǫglar veðr
léku við ský
of bauga.
Umðu
oddláar
í Óðins
veðri;
Hné mart
manna fyr
mækis
straumi.

Red colours mingled beneath the sky of the shield-rim [SHIELD]; the storms of Skǫgul <valkyrie> [BATTLE] played against the clouds of shield-rings [SHIELDS]. Point-waves [BLOOD] roared in the storm of Óðinn [BATTLE]; many people sank down before the tide of the sword [BLOOD].
(SKALD 1: 183)

In this example, we revisit the previously discussed Old Norse parallel between the chaos of a storm and the battlefield, as the kenning 'storm of Óðinn' is used in *Hákonarmál* to represent battle. The kenning fits into a common kenning pattern that is found in praise poems, which is that <weather> of <non-human being> equals battle.[7] All the same, *Óðins veðr* cannot be taken as a literal reference to weather magic, as kennings are generally metaphorical. In this case, the poet's intended meaning of the kenning *Óðins veðr* contributes to what Hannah Burrows (2021) describes as the implied meaning of a stanza, or poem, once the kennings are resolved. However, as Burrows suggests, the poetic language used to formulate kennings can also create a second layer of meaning, which forms a separate, overarching imagery. In the case of *Hákonarmál*'s eighth stanza, the six kennings all reference the field of battle, Eyvindr's choice of poetic language constructing the image of a rising storm at sea (Burrows 2021: 70). In this secondary image, *Óðins veðr* gives the audience a sense that Óðinn plays a role in controlling the storm. *Óðins veðr* is Eyvindr skáldaspillir's only kenning that produces the image of Óðinn manipulating the weather, and control over the non-literal storm, or literal battlefield. Here, this ability is not limited to Óðinn, however. In the same stanza, a second weather-associated kenning suggests Skǫgul's control over the weather in the kenning *Skǫglar veðr* ('storm of Skǫgul'), also a kenning for 'battle'.

Other battle kennings in skaldic verse that utilise Óðinn's association with the weather to enhance the secondary image of an escalating storm. *Háleygjatal* (composed c. 985), a praise-poem that traces Hákon jarl Sigurðarson's ancestry back to Óðinn, also uses a storm produced by Óðinn as a reference for battle:

6. Ok sá halr
at Hǫ́ars
veðri
hǫsvan
serk
Hrísgrís
nis bar.

And that man wore the grey shirt of Hrísgrísnir <wolf> [WOLF-SKIN] in the storm of Hǫ́arr <=Óðinn> [BATTLE].

(SKALD 1: 204)

In this stanza, *Hǫ́ars veðr*, or 'the storm of Hǫ́arr <=Óðinn>' is used to refer to the battlefield, *Hǫ́ars veðr* being almost identical to the kenning employed in stanza 8 of *Hákonarmál*, except for the use of the name Hǫ́arr, an alternative appellation for Óðinn. A similar kenning is used in the tenth stanza of *Vellekla*, a skaldic poem attributed to Einar skálaglamm Helgason. It is difficult to be certain of *Vellekla*'s dating, but previous scholarship (see SKALD 1: 282) suggests that the poem was composed during Jarl Hákon's lifetime, sometime before c. 995. In the poem's depiction of Jarl Hákon's violent conflict with the sons of Eiríkr blóðøx, Einar employs a kenning in which the snowstorm of Hǫ́arr (Óðinn) is used to describe the escalation of the battle:

10. Rignði hjǫrs á
hersa ok hald-
Viðurr haulda
hriðremmis
fjǫr víða
haffaxa lét vaxa

– þrimlundr of jók
Þundi Laufa veðr at
lífi
þegns gnótt –
méilregni lifkǫld
Hǫars drífu.

The arrow-rain [BATTLE] of the strengthener of the storm of the sword [(lit. 'storm-strengthener of the sword') BATTLE > WARRIOR] rained widely on the life of the *hersar*; the battle-minded one increased the abundance of retainers for Þundr <= Óðinn>. And the steering of Viðurr <=Óðinn> of sea-horses [SHIPS > SEAFARER] let the life-cold storms of Laufi <sword> [BATTLES] grow against the life of men in the snow-storm of Hárr <= Óðinn> [BATTLE] (SKALD 1: 295).

The kenning *Hǫars drífr* is an interesting choice of language in that it suggests Óðinn has the ability to manipulate wintry weather, rather than his more common control over wind at sea. Burrows (2021:74) nonetheless suggests that Óðinn's role in the construction of this meteorological kenning is perhaps unsurprising, due to his relationship with Jarl Hákon as a god with warrior attributes, and his relationship with the poem's poet, Einarr, as a god skilled in poetry. I would go even further in this interpretation by suggesting that Einarr was aware of Óðinn's reputation of having skills in magic and was deliberately employing the god's aptitude for weather manipulation in the construction of the entire stanza. This, to my mind, is evident because alongside the use of *Hǫars drífr*, the tenth stanza of *Vellekla* also references Óðinn's two weather-related names, Viðrir and Þundr.

Óðinn's Weather *Heiti*

Viðrir

Óðinn is associated with over two hundred *heiti*, or appellations, which appear throughout the Old Norse corpus and provide insight into the different personal qualities that were attributed to the god. The highest percentage of names given to Óðinn have less to do with his magical capabilities and more with his association with war and aggression.[8] Nevertheless, an examination of the long list of his differing appellations tells us much about his knowledge in magic. As previously explained, battle scenes and harsh weather are inherently connected in the Old Norse corpus, so it makes sense that a name connected to Óðinn's control over the weather might also be a reference to the god's aggressive, war-like nature. This approach, however, is by no means flawless, as there are often differing ways of interpreting the meanings of Óðinn's *heiti*. Nonetheless, an analysis of the surrounding context that accompanies the use of these names provides a little more insight into how weather magic is attributed to Óðinn in the sources.

The most frequently used of the two weather-related names of Óðinn is Viðrir,[9] often glossed in modern scholarship as 'weather-god' or 'he who rules the weather' (Price 2019: 82). Hjalmar Falk (1924: 34) connects this name directly to Óðinn's ability to magically manipulate the weather, the reoccurrence of this name in poetry perhaps suggesting that Óðinn's reputation as a god skilled in weather magic was more commonly acknowledged than scholarship has previously considered. Although this *heiti* is found within a range of Old Norse sources, including skaldic poetry, Eddic poetry, and prose, it is only in skaldic poetry that the name is invoked in the context of clear imagery of weather and climatic manipulation.[10]

Viðrir is employed twice in Tindr Hallkelsson's *Hákonardrápa* (c. 985), which is another poem praising Hákon jarl Sigurðarson. The name first appears in the third stanza, during the poem's portrayal of the escalating battle of Hjǫrungavágr:

3. Dreif at
Viðris veðr
þars í sundr
á sandi
– vargr gleyþti ná margan –
Sǫrla blés fyr jarli
– varð auðfundit
virð*i (þess hefr
seggja sessi)
valgagls – þinurs
hagli, serk
hringofinn
(merki).

The hail of the bow [ARROWS] pelted in the storm of Viðrir <= Óðinn> [BATTLE] – the wolf swallowed many a corpse; the meal of the slaughter-goose [RAVEN/EAGLE > CORPSES] was easily found –, where the ring-woven shirt of Sǫrli <legendary hero> [MAIL-SHIRT] was blasted apart on the sand before the jarl; the benchmate of men [RULER = Hákon] bears the marks of this.

(SKALD 1: 343)

Here, *Viðris veðr*, or the storm of Viðrir, is used as a stand-in for 'battle', similar to the use of *Óðins veðr* in in *Hákonarmál* (st. 8) and *Hǫars veðr* in

Háleygjatal (st. 6). Later in *Hákonardrápa*, Tindr uses Viðrir in a similar pattern, once again to invoke chaos on the battlefield:

10. Þás fyr borð
Mikinn gerði herr hjǫrva
* bǫðvar fangs at
ganga hljóm*; runnu
þá sunnan
veðrmagnanda
Viðris bauga skeiðr
at breiðu virðendr
Búa kenndu.
Baldrs Hǫ́kunar valdi.

When the guardians of the tunic of battle [MAIL-SHIRT > WARRIORS] on the ship taught the strengthener of the storm of Viðrir <= Óðinn> [lit. 'storm-strengthener of Viðrir') BATTLE > WARRIOR], Búi, to go overboard. The army made a great tumult of swords [BATTLE]; the warships of the Baldr <god> of rings [MAN] then ran from the south towards the wide domain of Hákon.

(SKALD 1: 354)

Like many of the poems depicting Hákon jarl Sigurðarson's efforts during the battle of Hjǫrungavágr, *Hákonardrápa* employs weather imagery throughout the poem as a means of enhancing and complimenting the chaos and destruction brought about by battle,[11] although here the allusions to weather are not limited to depictions of weather manipulation. In this poem, however, it is noteworthy that Óðinn is the only god or non-human to be used in the <weather> of <non-human being> kenning pattern, underlining that Óðinn is repeatedly invoked as a figure through whom weather imagery is mobilised (Burrows 2021: 69). The kenning *veðrmagnandi Viðris*, or 'the strengthener of the storm of Óðinn', creates a secondary image of Hákon being able to strengthen the weather that Óðinn created, highlighting Hákon's heroic efforts on the battlefield. Another instance of humans controlling the weather is the phrase *veðreggjendr* or weather-inciters, which appears in the sixth stanza, which relates the close of the battle of Hjǫrungavágr:

6. *Gjǫrðusk Gǫndlar borða*
*hǫggva glaumr óx þar at * Nauma –*
auði grimms at eyðask
ǫll lǫnd Dana brandi.
Kennt hefr heggr at
hræ birtinga sævar*
veðreggjǫndum viggja
veggs niðr of þat skeggi.

All the lands of the one hostile to riches [Hákon] came to be laid waste with the sword of the Danes; the revelry of the planks of Gǫndul <valkyrie> [SHIELDS > BATTLE] increased there at Nauma. The bird-cherry of the corpse-trout [SWORD > WARRIOR = Hákon] has taught the inciters of the weather of the wall of the horses of the sea [(lit. 'weather-inciters of the wall of the horses of the sea') SHIP > SHIELD > BATTLE > WARRIORS] to hit down with their bears on account of that.

(SKALD 1: 348)

In both *veðreggjendr* and *veðrmagnandi Viðris*, the implication that humans have some control over Óðinn's weather is used to enhance the imagery of the battle's chaotic destruction, or stress Hákon's military superiority, rather than actual weather manipulation (Burrows 2021: 73). However, in the same way that the poetic language of the <weather> of <non-human being> kenning pattern creates a second layer of meaning in the form of a weatherscape, the two kennings that are constructed from humans controlling the storm perpetuates the poetic image of weather-magic practitioners (i.e. the warriors) controlling the storm (i.e. the battle's) outcome.

Þundr

The second of Óðinn's weather-related names is 'Þundr'. Falk (1924: 31) classifies this *heiti* as one that is very old, belonging to the Germanic verb *Þindan*, meaning 'swell'. Falk (1924: 31) interprets the use of Þundr as evidence that Óðinn was known for having control over the restless seas. Price (2019: 65) nonetheless associates the name's evolution from 'swell' as an indication of Óðinn's anger, rather than being associated with his skills in weather magic. Although I agree with Price that 'swell' does invoke a sense of building anger, I would also suggest that Falk's proposal that Þundr is evidence of Óðinn's ability to control the seas is still valid. The appropriateness of connecting the name directly to Óðinn's knowledge of weather magic is demonstrated through the use of Þundr in *Vellekla*, where it appears in the context of poetic language that links Óðinn to the weather. The classifications of Óðinn's names into single categories can be problematic in that Óðinn's differing skills are often intertwined with one another. Indeed, the inclusion of the name Þundr within stanza 10 of *Vellekla*, noted above, creates a strong picture of weather imagery, and alongside the kenning '*Hǫars drífr*', once again demonstrates that Óðinn's ability to control the weather seems to be inherently linked to his reputation for aggression on the battlefield. In this stanza, all of the kennings refer to what is happening on the battlefield, the overarching poetic imagery creating the image of a chaotic storm at sea.

In addition to its appearance in *Vellekla*, Þundr is also used in Sturla Þórðarson's *Hrafnsmál* (composed after c. 1263) which details Hákon Hákonarsson's campaign in Scotland, aiming to re-establish his control over the Hebrides. This poem is thought to have been composed following the death of Hákon Hákonarsson in 1263, so it is not a direct representation of Óðinn from the Viking Age. It is, however, interesting that it uses one of Óðinn's names that is connected to weather manipulation and the violence in battle:

20. Beit at
brynmóti
brún í
rauðtúnum
egg of
ódyggvar
aldar
gunntjalda,
áðr en
egghríðar
undan
víglundum
skyndu ský-
Þundar
skoskir
alþroskins.

The shining sword-edge bit in the red enclosures of battle-tents [SHIELDS > SHIELD-WALLS] around unfaithful men at the byrnie-meeting [BATTLE], before the Scottish Þundar <= [Óðinns]> of the cloud of the edge-storm [(lit. 'cloud-Þundar of the edge-storm') BATTLE > SHIELD > WARRIORS] hastened away from the slaughter-trees [WARRIORS] of the very manly one.

(SKALD 2: 744–745)

In this stanza, Þundr appears in the kenning *ský-Þundar*, meaning 'cloud-Óðinns'. This kenning appears as part of the phrase 'cloud-Þundar of the edge-storm [BATTLE > SHIELD > WARRIORS]', signifying the retreat of the Scottish warriors. Sturla Þórðarson's use of Þundr alone does not provide much evidence in terms of suggesting that Óðinn was still seen as having a connection to weather magic in the thirteenth century, outside of using the name alongside weather imagery. However, it is interesting to note that the poem uses the kenning *ský-Þundar skoskir*' to refer to the Scottish warriors, while also summoning an image of weather. Earlier in the poem, the Norwegian fleet is trapped near the Cumbraes in the Firth of Clyde by a storm that it is suggested may have been raised by Scottish weather magic:[12]

15. Mætti
margfréttin
n miklum
framstiklir
auðar
þjóða
inga
gjörningum
Leysti
lábrostinn
Lögr in
skautfögru
flaust ór
frónlæstum
festum
bauglestis.

The very inquisitive dispenser of wealth [GENEROUS MAN] encountered the mighty [magic] of the king of evildoers [Alexander]. The wave-bursting sea loosened the sail-fair ships of the ring-wounder [GENEROUS MAN] from the land-secured moorings.

(SKALD 2: 740–741)

16. Blés of böðfúsa
bragna
fjölmagnat
hregg á
herbygðum
hríðar
brimskíðum.
Upp rak
allkeppna öld
með
herskjöldu
grundar
gjálfrs undit
grand at
Skotlandi.

The [magic]-strengthened hailstorm blew around the battle-eager men on the troop-inhabited surf-skis of the sea [SHIPS]. The twisted harm of the land of the surge [SEA > STORM] drove the very vigorous crew with war-shields towards [the shores of] Scotland.

(SKALD 2: 740–741)

While the link between Óðinn and weather magic is not explicitly made in this poem, it is interesting that the poet includes episodes of weather magic, as well as using one of Óðinn's names in a kenning that invokes the image of control over the weather at sea by some people. As noted earlier, in the poem's fifteenth stanza, the magical storm is attributed to the Scottish king, Alexander. Schjødt (2008: 220) notes that although Óðinn is portrayed in textual sources as a master of *seiðr*, there is a lack of evidence suggesting that Óðinn is connected to general practitioners of magic. Schjødt gives support for this idea by suggesting that Óðinn's knowledge in magic must instead stem from the necessity in a society whose worldview is inherently tied to the paranormal, for a king or warrior to be versed in magic. Bearing this in mind, the allegation of Alexander's performance of weather magic, alongside the use of Óðinn's weather-related name when describing Scottish warriors, arguably provides support for Schjødt's idea of a longstanding connection existing between Óðinn's mastery of weather magic and the importance of magic to the elite and battle-focused members of society.

This relationship between Óðinn's weather magic and militaristic rulers and heroes is further demonstrated in Þorfinnr munnr's *Lausavísa* 2, in which the kenning *hregg Þundar*, or 'storm of Þundr' once again represents battle. The *lausavísa* is one of three stanzas attributed to Óláfr Haraldsson's skalds, which depicts the build-up of anticipation preceding the battle of Stiklastaðir (c. 1030).

2. Røkkr at regni
miklu randar
garðrs ins harða;
vill við visa
snjallan Verdœla
lið berjask.
Verjum allvald
ǫrvan; ǫlum
teitan mǫ́ sveita;
fell Þrœndr í
Þundar
Þess eggjumk vér – hreggi.

It grows dim approaching the great downpour of the tough enclosure of the shield-rim [SHIELD > BATTLE]; the force of the Verdœlir wants to fight against the valiant prince [Óláfr]. Let us defend the generous mighty ruler; let us feed the cheerful goal of the gore [RAVEN/EAGLE]; let us fell the Þrœndr in the storm of Þundr <= Óðinn> [BATTLE]; we are urging this.

(SKALD 1: 847)

As can be seen here, this stanza uses 'storm of Þundr' in the earlier-noted <weather> of <non-human being> kenning pattern, alongside other weather imagery, to illustrate the chaos of battle, for example, the weather kennings from *Vellekla* and *Hrafnsmál*. Although this example goes hand in hand with Price's classification of Þundr in the aggression category and Schjødt's suggestion that Óðinn's magic is connected to warriors, examples also exist of *Þundr* being used in kennings that seemingly represent notions entirely unrelated to the battlefield. In the *drápa Rekstefja*,[13] the kenning *Þundregn*, or 'Þundr's rain', is used to refer to poetry:

8. Haukjóðs harða
víð (hǫtt)
norrœnar
dróttir*
(Þundregns þeim of
vandak) þengils á bý
gengu.
Óráð illri
þjóðu Óláfr
of galt dála;
vígrunnr
velja kunni
víkingum hlut slíkan.

Norwegian troops placed themselves very widely in {the homestead {of the hawk-offspring}} [HAWK > HAND] of the ruler; I fashion for him the metre {of Þundr's <= Óðinn's> rain} [POETRY]. Óláfr thoroughly paid evil people for their misdeeds; {the battle-tree} [WARRIOR] was able to deal vikings such a fate.

(SKALD 1: 905)

The line of the poem that includes the *Þundregn* kenning explains that the poet has fashioned this poetry for a ruler. In this case, the <weather> of <non-human being> kenning pattern is used as a stand-in for poetry (SKALD 1: 905), rather than a means to illustrate the chaos of the battlefield, albeit poetry composed under circumstances that relate to a battle: prior to the Þundr-kenning's use, the poet describes a scenario involving Norwegian troops. The poem as a whole is constructed to portray a scene involving conflict, so it makes sense that the poet should choose Óðinn's name to construct the kenning, as he is often said to be the controller of the outcome of a battle.

Conclusions

As noted at the start of this chapter, modern reconstructions of Norse mythology and religion often rely on scholars piecing together fragmented poetry and other mythological evidence that is preserved in textual sources, all written down generations after the Scandinavian conversion to Christianity. This inevitably leaves gaps in these cultural reconstructions, making it almost impossible to establish a completely genuine representation of Viking Age myths and religion. Nevertheless, the sheer number of references that associate Óðinn with weather control suggests strong literary and cultural associations between Óðinn and weather magic in the time in which the sources were written down.

The conclusion that Óðinn is repeatedly represented as a master of weather magic, primarily focused on changing the winds or subduing storms at sea, challenges previous arguments that Óðinn was not particularly relevant to sailors. As Terry Gunnell (2017: 122) suggests, Óðinn had seemingly little importance in Iceland outside the court poets' sphere, with only a limited number of Icelanders engaging with the god's skillsets as portrayed in the surviving texts. Few Icelanders belonged to the great Danish and Norwegian armies (Gunnell 2017: 122), and previous scholarship suggests that Óðinn would have had little relevance for farmers, traders, and sailors. However, as evidenced throughout this chapter, Óðinn's weather magic is consistently depicted in sailing scenarios within skaldic verse involving kings and warriors. There are no existing skaldic examples of Óðinn performing weather magic to aid general sailors or anyone without a militaristic motive, which might further substantiate Schjødt's (2008: 220) suggestion that Óðinn's skills in magic stem from the ruling elite's reliance on magic to rule successfully.

Óðinn's portrayal as a weather-magic practitioner underscores his role as a deeply complex and multi-faceted deity. His attributes and capabilities extend far beyond war and aggression, encompassing a wide range of roles, including poetry, seduction, and magic. This multifaceted nature is evident in the skaldic poetry and saga literature reviewed in this chapter where Óðinn's roles in the texts frequently overlap and intertwine. For example, in the Eddic poem *Hávamál*, Óðinn's mastery of magic is showcased through a list of spells that include the ability to calm the seas, underscoring his profound wisdom and esoteric knowledge. Óðinn's association with weather magic is particularly compelling and vividly portrayed in texts such as *Ynglinga saga* and *Reginsmál*. In *Ynglinga saga*, Óðinn is described as having the power to calm the sea and control the winds with words alone. Similarly, in *Reginsmál*, he calms a storm encountered by Sigurðr and his men, firmly establishing him as a master of weather magic. These accounts highlight Óðinn's profound influence over the elements, particularly in maritime contexts.

The connection between Óðinn's weather magic and warfare is another significant theme that has been highlighted in this chapter. In Old Norse literature, weather-related kennings often describe battles, drawing a parallel between the chaos of a storm and the turmoil of combat. Skaldic poetry employs phrases like *Óðins veðr* (storm of Óðinn) to represent battle scenes, reinforcing the notion that Óðinn's weather manipulation is linked to his warlike nature and the chaotic forces of nature and conflict. This association is evident in poems like Tindr Hallkelsson's *Hákonardrápa*, which praises Hákon jarl Sigurðarson and uses weather-related kennings to depict the chaos and destruction of battle.

Óðinn's weather magic is shown to be particularly relevant to the elite and warrior classes. These depictions highlight the importance of Óðinn's weather-manipulating abilities in enhancing the heroic imagery associated with kings and warriors. The recurrent theme of weather magic in such contexts brings to light its significance in maintaining power and achieving military success. Despite the prominence of Óðinn's weather magic in elite and warrior contexts, there are few references to its use by common sailors. This suggests that Óðinn's abilities were primarily valued for their strategic and symbolic significance in warfare rather than for practical navigation by everyday seafarers.

A demonstrative example is found in the fifth chapter of *Hallfreðar saga vandræðaskálds*, where the crew contemplates offering money to Freyr, Óðinn, or Þórr based on the wind's direction. Although this scene highlights the gods' influence over navigation, it is still connected to a skaldic poet and retains a link to court culture and elite concerns (CSI 1: 233). While this scene perhaps links Óðinn's weather magic to a slightly different non-militaristic sailing scenario, it is important to keep in mind that the poem itself stems from a skaldic poet, reflecting Óðinn's significant role in court culture. Skaldic poetry provides substantial evidence of Óðinn's association with weather magic. The intricate use of kennings, such as 'Óðins veðri' and 'Viðris veðr', to describe battles, illustrates the cultural and literary significance of Óðinn's weather-manipulation abilities. These kennings, often used to enhance the depiction of chaotic battle scenes, reflect the broader belief in Óðinn's control over the weather. In the poem *Vellekla*, for example, the use of the kenning 'Hǫars drífu' (snow-storm of Hǫarr) suggests that Óðinn's influence extends to wintry weather conditions as well, further solidifying his role as a weather-magic practitioner.

The broader circulation of Óðinn's weather magic in the Scandinavian Middle Ages is another significant finding in this chapter. Numerous references across different genres of Old Norse literature indicate that the connection between Óðinn and his ability to magically manipulate the weather was widespread throughout the Scandinavian Middle Ages. This is evidenced by the consistent depiction of Óðinn calming storms in texts like *Reginsmál* and *Ynglinga saga*, as well as in texts such as *Gesta Danorum*. Additionally, the use of weather-related kennings in skaldic poetry further illustrates the prevalence of this association. These references collectively suggest that the connection between Óðinn and weather magic was more commonly acknowledged and integrated into Old Norse literature than previously considered.

In conclusion, the study of Óðinn's reputation as a weather-magic practitioner reveals another aspect of his character within Norse mythology, adding to the many magical skills that have already been tied to the god throughout the extant textual sources. Óðinn's multifaceted nature as it is portrayed in literature, encompassing war, poetry, seduction, and magic, is intricately linked to his ability to control the weather. This association is consistently portrayed in various Old Norse texts, highlighting the god's profound influence over the natural world, particularly in maritime contexts. The connection between weather magic and warfare underscores the chaotic and destructive aspects of both, reinforcing Óðinn's warlike nature and his significance in elite and warrior culture.

The widespread circulation of Óðinn's weather magic in the Scandinavia Middle Ages is evidenced by numerous references across different genres of Old Norse literature. This comprehensive analysis of the available source material enhances scholars' understanding of Norse mythology by shedding light on the interconnectedness of weather magic, warfare, and elite culture in medieval Scandinavian society. By examining the various literary and poetic sources, this chapter provides a detailed exploration of Óðinn's weather-magic capabilities and his cultural significance during the period in which the sources were written down. It reinforces the importance of considering the fluidity and dynamic nature of divine roles in Norse mythology, as they are portrayed in literature, when reconstructing historical mythologies.

Notes

1 See Schjødt 2019 for a survey of the recent scholarship concerning the historical development of Óðinn.

2 This tale about memories of ancient times is preserved in the manuscript AM 62 fol., which is normally dated between c. 1375 and 1400 (Ólafur Halldórsson 2000).

3 This legendary saga is preserved in the manuscript NKS 1824 b 4to, which is normally dated between c. 1380-1420 (Byock 1990).

4 '*Þeir viku at landi ok tóki karl á skip sín. Þá tók af veðrit, ok fara unz þeir koma at landi í ríki Hundings sona. Þá hvarf Fjǫlnir*' (Finch 1965: 29) ['They put to land and took the man on board. Then the storm abated, and they sailed until they reached land in the kingdom of Hunding's sons. Then Fjolnir vanished' (Finch 1965: 30)].

5 Óðinn gifts Starkaðr a life three times as long as a normal lifespan, possession of the finest weapons, an abundance of wealth, victory on the battlefield, a talent for composing poetry, and being held in high regard by the most rich and influential members of society. Þórr, on the other hand, curses Starkarðr with infertility, the fate of committing a large crime in each of his three lifetimes, an insatiable appetite for wealth and property, always receiving a substantial wound in every battle,

to never being able to remember his own skaldic poems, and being hated by the common people (FSN 4: 28-31).

6 Unlike in the episode in *Gesta Danorum*, Víkarr is now killed by a fake noose suddenly turning into real rope, which seems to be the result of Óðinn's intervention.

7 The <weather> of <non-human being> was first highlighted in Hannah Burrows's (2021) presentation on 'weatherscapes' in skaldic poetry. For the modern classifications of extant kennings, see Meissner 1921.

8 For a full list of Oðinn's names and discussion of their magical implications, see Neil Price's *The Viking Way* (2019); for a philological breakdown of Oðinn's names, see Hjalmar Falk's *Odensheite* (1924).

9 Alongside Hárr and Þriði (Price 2019: 82).

10 For example, in stanza 26 of the eddic poem *Lokasenna*, Óðinn is referred to as Viðrir when Loki shames Frigg, Óðinn's wife, for sleeping with Óðinn's brothers. Outside of the use of the name itself, there are no further connections between Óðinn and the weather in *Lokasenna*. Viðrir also appears in *Gylfaginning* where the name is included in a list of various names that Óðinn calls himself on a visit to King Geirrøðr. Once again, the name appears without any further detail or narrative context being given that would suggest Óðinn's skills in weather magic.

11 See Burrows (2021) for a discussion on the use of 'weatherscapes' in skaldic verse.

12 Fulk highlights the fact that the storm is also mentioned in Insular sources, but it is attributed to divine intervention rather than a magical performance (SKALD 2: 740). See Anderson 1922, 2: 607 for more discussion of these sources (SKALD 2: 740).

13 This poem is attributed to Hallar-Steinn, about whom we know little. The poem's praise of Óláfr Tryggvason might suggest that the poem was composed in Iceland toward the end of the twelfth century (SKALD 1: 893).

CONCLUSION

The Cultural Climate of the North

In conclusion, this study provides a comprehensive exploration of the cultural significance and societal perceptions of weather magic in medieval Scandinavia, covering the period from approximately 800 to 1555 AD. Through an interdisciplinary approach, this study has examined the portrayal of weather magic in Old Norse sagas and related materials, with the aim of deepening our understanding of its narrative functions, societal relevance, and cultural implications. By integrating textual analysis, archaeological perspectives, and cultural insights, this research has illuminated the multifaceted role of weather magic within medieval Scandinavian society, highlighting its profound impact on various aspects of life, including ritual practices and performances, social dynamics, and power structures. The following conclusion ties together the key findings and implications of this study, offering a nuanced interpretation of the enduring legacy of weather magic in shaping historical narratives and cultural practices.

Summary of Key Findings and Implications of Study

By exploring the prevalence and cultural significance of weather magic in medieval Scandinavia, this research contributes significantly to our understanding of medieval Scandinavian history and offers new insights into the pivotal role of magic in shaping historical narratives and events within the region. Weather magic, as portrayed in sagas, poems, and other textual sources, serves as a lens through which we can examine the complex relationship between medieval Scandinavians and their environment.

First and foremost, this research project illuminated how weather magic was deeply rooted in the culture and religion of medieval Scandinavian society.

DOI: 10.4324/9781003746058-10

By analysing the depiction of weather magic in Old Norse sagas and other literary works, this study revealed the ways in which medieval Scandinavians perceived and interpreted natural phenomena. Weather magic reflects medieval attempts to make sense of the unpredictable forces of nature and to exert a measure of control over their environment. Moreover, the portrayal of weather magic in these texts underscores its cultural significance as a tool for both practical survival and cultural expression within medieval Scandinavian society.

The overarching key findings of this study highlight the profound integration of weather magic into the cultural, social, and religious aspects of medieval Scandinavian society. This study revealed that weather magic was not merely a literary device but a deeply rooted belief system that was integrated into various aspects of medieval Scandinavian life, influencing everything from everyday survival to social dynamics and power structures.

Weather magic endowed characters with significant agency, affecting narrative outcomes and shaping societal perceptions. The performative nature of weather magic, with its reliance on vocal and symbolic actions, highlighted the critical role of ritual performance and tradition in these practices. The intricate connection between weather magic and broader magical traditions, such as *seiðr*, demonstrates the importance of these weather-magic acts in the cultural environment. Vocal performances, incantations, and symbolic gestures were not only central to the execution of weather magic but also to its recognition, acceptance, and perpetuation within medieval Scandinavian society.

Furthermore, the study demonstrated the fluidity and adaptability of religious practices during this period. Christian authorities often integrated and reinterpreted non-Christian magical elements into their religious frameworks. This synthesis of practices illustrates how Christian figures pragmatically adopted elements of pre-Christian traditions to address local spiritual and communal needs. The blending of prayer, holy water, and other Christian rites with non-Christian weather-magic practices reflects a dynamic and adaptive approach to religious and cultural transformation.

The exploration of Indigenous weather magic highlighted the cultural exchanges and mutual influences between Norse and Indigenous communities, particularly the Saami. This examination revealed the Norse acknowledgment of and reliance on Indigenous magical skills and suggested a dynamic exchange of cultural practices. The recognition and regulation of these intercultural exchanges in legal texts further illustrate the interconnectedness of Norse and Indigenous traditions. The portrayal of Indigenous weather magic in medieval Scandinavian texts underscores the necessity of avoiding oversimplification and recognising the diversity within these practices.

Moreover, the present study sheds light on the social dynamics and power structures that surrounded the practice of weather magic in medieval

Scandinavia. Through close examination of the roles and identities of weather-magic practitioners, as well as the consequences of their actions, scholars can gain insights into broader issues of authority, identity, gender, and social order within medieval Scandinavian society. Weather magic, in particular, often intersected with themes of conflict, revenge, and political manoeuvring in the sagas, highlighting its multifaceted role in shaping historical narratives and interpersonal relationships.

The book also highlighted the role of weather magic in reinforcing social hierarchies and power structures. The association of weather magic with elite figures and its depiction in the context of warfare and maritime activities underscored its significance for the military elite. This connection between magical practices and social power highlighted the strategic importance of weather magic in legitimising and elevating the military elite, aligning them with divine power and favour.

Furthermore, weather magic was not confined to Scandinavia alone during the Middle Ages. Examples of weather magic can be found across continental Europe during the same period. This widespread practice underscores the universal importance of weather to human existence, particularly in agrarian and seafaring societies where livelihoods depended heavily on successful harvests and favourable sailing weather conditions. Whether in Scandinavia or elsewhere in Europe, the use of weather magic reflects the deep-seated belief in human agency over the natural world and the desire to mitigate the uncertainties of life through magical means.

The interdisciplinary approach of this study, which integrated textual analysis with archaeological and anthropological perspectives, underscored the interconnectedness of medieval Scandinavian cultures and the shared human endeavour to control and understand the natural world. The careful examination of weather magic in various contexts, from ritual practices to legal regulations, revealed a complex system of beliefs and practices central to medieval Scandinavian life. This approach provided a comprehensive understanding of how weather magic functioned within and between these communities.

By drawing upon a diverse array of sources, the approach used in this study transcends traditional literary groupings, offering a holistic and multifaceted perspective on the phenomenon of weather magic in medieval Scandinavia. The inclusion of a variety of textual sources, such as Old Norse sagas, poems, and legal documents, provides insights into the literary representations and cultural perceptions of weather magic, offering insight into how it was understood and potentially practised by medieval Scandinavians. Furthermore, although the study argues against the existence of material evidence for weather-magic practices in medieval Scandinavia, the incorporation of discussions surrounding archaeological evidence that has previously been presented as footprints of weather-magic practices reinforces the significance of vocal performance in weather-magic practices during the time period.

The study's engagement with legal sources offers a unique perspective on the social and legal frameworks that governed magical practices in medieval Scandinavia. By analysing legal codes and witch trials related to general magic and weather magic itself, this study illuminates the ways in which magical practices were regulated, contested, and sanctioned within Scandinavian society. The incorporation of linguistic analysis deepens our understanding of the language and terminology associated with medieval Scandinavian weather magic. By examining the etymology, semantics, and usage of magical terms and phrases in medieval Scandinavian texts, this study unveils the linguistic nuances and cultural connotations embedded within magical discourse during the time period.

The findings of this study contribute to a deeper understanding of medieval Scandinavian history, offering valuable insights into the complexities of human existence and the intricate interconnectedness of magic, religion, and culture during this period. The enduring legacy of weather magic continues to shape contemporary interpretations of medieval narratives and cultural practices. The portrayal of weather magic in both everyday life and elite culture demonstrates its significance across all social strata.

In conclusion, this book has illuminated the multifaceted role of weather magic in shaping social dynamics, reinforcing power structures, and fostering cultural exchange. The study has underscored the importance of a nuanced and interdisciplinary approach to understanding the rich and diverse magical traditions of medieval Scandinavia. By examining the cultural, social, and religious contexts in which weather magic was presented, the study has provided a comprehensive and carefully considered interpretation of its historical and cultural legacy. The insights gained from this study contribute to a more holistic understanding of medieval Scandinavian society and the enduring impact of its magical traditions on historical and cultural narratives.

Future Directions for the Study of Scandinavian Weather Magic

Future research could benefit from an investigation of gender dynamics in the practice and perception of weather magic. While this book has touched upon the roles of various practitioners, a more focused study on how gender influenced the practice, transmission, and societal views of weather magic could provide deeper insights. Exploring how the different gender identities of the practitioners were portrayed differently in Old Norse sagas and other texts, and the implications of these portrayals on our understanding of gender roles in medieval Scandinavian society, would be valuable. Additionally, examining the intersection of gender with magical practices could uncover how societal expectations and power dynamics shaped the experiences and reputations of weather-magic practitioners. This line of inquiry could also illuminate the ways in which gender influenced the later accusations, trials, and punishments related to weather magic.

Another promising direction for future research is to employ a comparative approach, examining weather magic in medieval Scandinavia alongside similar practices in other cultures and regions. By comparing Scandinavian weather magic with practices in other European cultures, scholars could identify common themes and unique characteristics. This comparative analysis could reveal broader patterns in human interaction with the natural world and the role of magic across different societies. For instance, understanding how different cultures used rituals and performances to influence weather can provide deeper insights into the universal human desire to control natural forces and mitigate environmental uncertainties. Such studies would not only enhance our understanding of medieval Scandinavian weather magic but could also contribute to the broader field of comparative religion and anthropology by highlighting the shared human experiences and cultural expressions of weather magic.

A comparative study between Scottish and Scandinavian weather magic would provide a compelling framework for exploring the cultural and historical connections between these regions. Given the historical ties between Scotland and Scandinavia, particularly in regions like the Northern Isles, the Hebrides, and Caithness, an examination of the commonalities and differences in weather-magic practices would offer valuable insights into shared cultural influences and regional variations. The intersection of Scottish and Scandinavian weather magic is further realised in the early modern period, notably with the ascension of King James VI to the Scottish throne. The matrimonial union between James VI and Anne of Denmark, sister to Christian IV of Denmark, in 1589, precipitated a sequence of events that led to separate witch trials taking place in both countries that heavily featured accusations of weather magic.

After a proxy marriage to King James VI in Helsingør, Denmark, Queen Anne's voyage to Scotland was met by a series of ferocious storms, forcing the fleet to reroute to Oslo. These tempests were swiftly attributed to a cohort of Danish women accused of witchcraft, whose purported weather magic incited the Copenhagen witch trials of 1590. During this trial, six Danish women admitted guilt in conjuring the storms that rerouted Queen Anne's journey, two of whom were burned at the stake at Kronborg Castle. King James VI, prompted by the dangerous circumstances surrounding his consort's voyage and the trial in Denmark, instigated his own tribunal, centred in North Berwick, where individuals stood accused of conspiring to assassinate the king through the conjuration of storms.

An examination of these events from both Scandinavian and Scottish perspectives in a comparative study of weather magic could yield valuable insights into shared cultural influences and regional variations. By juxtaposing the reactions and interpretations of weather-related phenomena in both contexts, scholars would perhaps uncover differences in magical beliefs and societal responses. Moreover, such a comparative analysis could illuminate broader patterns of magical belief and social dynamics prevalent in medieval and early modern Europe.

Furthermore, the examination of the diffusion of Scandinavian weather-magic motifs into English discourse during the medieval period, particularly concerning the practice of wind knots, merits scholarly inquiry. A deeper examination of English sources that reference wind knots (introduced in Chapters 3 and 4 of the present study), irrespective of their attribution to Saami practitioners, could offer a foundational exploration into the transmission of these magical beliefs. A chronological exploration, spanning from these initial English sources to subsequent Scandinavian accounts by figures such as Olaus Magnus and Johannes Schefferus, would provide an opportunity for investigating the evolution of perceptions surrounding wind-knot magic. Of particular interest is discerning whether the medieval British depictions of wind-knot magic could have influenced Scandinavia's own conceptualisation of this phenomenon, potentially culminating in wind-knot accusations in witch trials within Finnmark.

Expanding the study of medieval Scandinavian weather magic into the early modern period presents an exciting opportunity to investigate the evolution of weather magic and the medieval influence on the Early Modern witch trials that took place all over Scandinavia. By tracing the continuity and transformation of medieval concepts of weather magic into the Early Modern period, scholars could gain insights into how magical beliefs persisted and adapted amidst changing social, religious, and cultural contexts. Investigating the connections between medieval Scandinavian weather magic and the accusations of weather manipulation in the Early Modern witch trials could offer a nuanced understanding of the cultural legacy and enduring significance of weather-magic beliefs in Scandinavian society.

In conclusion, the proposed future research directions highlight the potential for enriching our understanding of medieval Scandinavian weather magic and its broader implications. Comparative studies with other cultures, a deeper examination of gender dynamics, and extending the investigation into the early modern period are promising avenues that can provide fresh perspectives and nuanced insights. By continuing to explore these dimensions, scholars can further illuminate the complex interplay of magic, culture, and society in medieval and early modern Scandinavia and beyond.

Closing

In conclusion, the comprehensive examination of weather magic within medieval Scandinavian society reveals its significance as more than a mere literary motif. Rather, it serves as a reflection of the intricate cultural, social, and belief systems prevalent during the time period from 800 to 1555.

Through a careful analysis that integrates textual analysis, archaeological insights, and cultural perspectives, this study has illuminated the multifaceted role of weather magic in shaping various aspects of medieval Scandinavian culture, including literary traditions, religious thought, social dynamics, and power structures.

The study of weather magic extends beyond the boundaries of medieval Scandinavia, offering insights into broader historical and cultural contexts. By situating Scandinavian weather magic within the framework of European medieval society and religious thought, this research underscores the interconnectedness of cultures and the universal human endeavour to understand and navigate the natural world. The cultural significance of weather magic in medieval Scandinavia is highlighted not only through its portrayal in Old Norse sagas and a variety of other medieval texts but also in its societal impacts, demonstrating an integration into the daily lives and beliefs of medieval Scandinavians.

Weather magic, as explored in this book, reveals a complex system of beliefs that intersected with broader magical traditions, such as *seiðr*. The detailed examination of terminology and practices associated with weather magic, such as vocal incantations, symbolic gestures, and the use of runes, showcases the deep cultural roots and the ceremonial nature of these magical performances. Weather magic was not an isolated phenomena but was deeply embedded in the social and religious thought of the time, reflecting a sophisticated understanding of the natural world and human agency within it in medieval Scandinavia.

The examination of Indigenous weather magic highlights the rich cultural exchanges and mutual influences between Norse and Indigenous communities during the medieval period. This interconnected portrayal of different cultures challenges simplistic dichotomies and emphasises the diversity within Indigenous practices, showcasing the dynamic and closely connected nature of cultural practices in Northern Fennoscandia. The recognition and regulation of these intercultural exchanges in legal texts further illustrate the complex relationship between Norse and Indigenous traditions.

Moreover, the study of weather magic in Christian and non-Christian contexts reveals the fluid boundaries and shared elements between these practices. The incorporation of non-Christian magical elements into Christian rituals, as seen in various Old Norse sagas, underscores the adaptive nature of religious practices during the medieval period. This combination of practices illustrates how traditions were continually reshaped to meet the changing needs and beliefs of society.

The findings of this study also highlight the role of weather magic in reinforcing social hierarchies and power structures. The association of weather magic with elite figures and its depiction in the context of warfare and maritime activities underscores its significance for the military elite. This connection between magical practices and social power highlights the strategic importance of weather magic in legitimising and elevating the military elite, aligning them with divine power and favour.

Furthermore, the study's approach to sources across a wide range of genres, which also integrates textual analysis with archaeological and cultural perspectives, emphasises the connections among medieval Scandinavian cultures and their collective human pursuit to dominate and interpret the natural world. The careful examination of weather magic in various contexts, from ritual practices to legal regulations, reveals a complex system of beliefs and practices central to medieval Scandinavian life.

The enduring legacy of weather magic continues to shape contemporary interpretations of medieval narratives and cultural practices. Ultimately, the comprehensive examination of weather magic within medieval Scandinavian society reveals its profound significance as a cultural, social, and religious phenomenon. This study not only enriches our understanding of medieval Scandinavian history but also offers valuable insights into the enduring human want to understand and influence the natural world. By situating weather magic within a broader historical and cultural context, this book underscores its integral role in shaping historical narratives and cultural practices, highlighting the rich and diverse magical traditions of medieval Scandinavia.

BIBLIOGRAPHY

Primary Sources

Adam of Bremen. 2002. *History of the Archbishops of Hamburg-Bremen*. trans. Francis J. Tschan. New York: Columbia University Press.

Agobard of Lyon. 1981. *Opera Omnia*, ed. L. Van Acker. Turnhout: Brepols.

AM 132 fol. (*Möðruvallabók*). c. 1330–1370. Stofnun Árna Magnússonar í íslenskum fræðum, Reykjavík.

AM 242 fol. (*Codex Wormianus*). c. 1350. Den Arnamagnæanske Samling and Stofnun Árna Magnússonar í íslenskum fræðum, Copenhagen and Reykjavík.

Babington, Churchill, ed. 1865–1866. *Polychronicon Ranulphi Higden monachi Cestrensis*, ed. Joseph Rawson Lumby. London: Longman, Green, Longman, Roberts and Green.

Bjarni Einarsson, ed. 1953. *Hallfreðar saga*. Copenhagen: J. Jørgensen.

Boer, R. C., ed. 1888. *Ǫrvar-Odds saga*. Leiden: Brill.

BS = Guðni Jónsson. 1953. *Byskupa Sögur*. 2 vols. Reykjavík: Íslendingasagnaútgáfan.

Byock, Jesse, trans. 1998. *The Saga of King Hrolf Kraki*. London: Penguin.

Cockayne, Thomas Oswald, ed. 1864. *Leechdoms, Wortcunning, and Starcraft of Early England: A Collection of Documents, for the Most Part Never Before Printed, Illustrating the History of Science in this Country Before the Norman Conquest*. London: Longman, Green, Longman, Roberts and Green.

Collinson, Lisa, Torgeir Landro and Bertil Nilsson, eds. 2020. *The Borgarthing Law and the Eidsivathing Law: The Laws of Eastern Norway*. Abingdon-on-Thames: Taylor & Francis.

CSI = Viðar Hreinsson, ed. 1997. *The Complete Sagas of Icelanders: Including 49 Tales*. 5 vols. Reykjavík: Leifur Eiríksson.

Dennis, Andrew, Peter Foote and Richard Perkins, trans. 2006. *Laws of Early Iceland: Grágás I. The Codex Regius of Grágás with Material from Other Manuscripts*. Winnipeg: University of Manitoba Press. (Reprint of 1980 edition.)

Detter, Ferdinand and Richard Heinzel, eds. 1903. *Sæmundar Edda: mit einem Anhang*. Vol. 2. Leipzig: G. Wigand.

Driscoll, M. J., ed. 1995. *Ágrip af Nóregskonungasögum = A Twelfth-Century Synoptic History of the Kings of Norway*. 2nd ed. London: Viking Society for Northern Research, University College London.

Dronke, Ursula, trans. 1997. *The Poetic Edda. Vol. II. Mythological Poems*. Oxford: Clarendon Press.
Dudo. 1865. *De moribus et actis primorum Normanniae ducum*. Caen: Le Blanc-Hardel.
Ekrem, Inger, Lars Boje Mortensen, and Peter Fisher, eds. 2003. *Historia Norwegie*. Copenhagen: Museum Tusculanum.
Emerton, Ephraim, trans. 1967. *The Correspondence of Pope Gregory VII: Selected Letters from the Registrum*. Vol. 14. London: Octagon Books.
Faulkes, Anthony, ed. 2016. *Færeyinga saga*. Dundee: Thorisdal.
Finch, R. G., ed. and trans. 1965. *The Saga of the Volsungs*. London: Thomas Nelson and Sons.
Finnur Jónsson, ed. 1905–1922. *Rímnasafn: samling af de ældste islandske rimer*. 2 vols. Copenhagen: S. L. Møllers bogtrykkeri.
Flowers, Stephen E., trans. 1989. *The Galdrabók: An Icelandic Grimoire*. York Beach, Maine: Samuel Weiser, Inc.
FSN = Rafn, Carl Christian, ed. 1829–1830. *Fornaldar sögur Norðrlanda*. 3 vols. Copenhagen: Ennu Popsku.
GKS 1005 fol. (*Flateyjarbók*). c. 1387–1394. *Stofnun Árna Magnússonar í íslenskum fræðum*, Reykjavík.
GKS 2367 4to (Codex Regius). c. 1300–1325. *Stofnun Árna Magnússonar í íslenskum fræðum*, Reykjavík.
Grágás I = Finsen, V., ed. 1974. *Grágás: Konungsbók*. Odense: Universitetsforlaget.
Gregory the Great. 1959. *Saint Gregory the Great: Dialogues*. trans. Odo John Zimmerman. Washington DC: Catholic University of America Press.
Guðni Jónsson and Bjarni Vilhjálmsson, eds. 1943–1944. *Völsunga saga*. Reykjavík: Bókaútgáfan Forni.
Guðni Jónsson, ed. 1954. *Fornaldar sögur Norðurlanda*. 3 vols. Reykjavík: Íslendingasagnaútgáfan.
Hast, Sture, ed. 1960. *Harðar saga ok Hólmverja*. Copenhagen: Munksgaard.
Haydon, F. S., ed. 1860. *Eulogium historiarum*. London: Longman, Green, Longman, Roberts and Green.
Heimskringla = Finlay, Alison, and Anthony Faulkes, trans. 2016. *Heimskringla*. 3 vols. London: Viking Society for Northern Research, University College London.
Hermann Pálsson and Paul Edwards, trans. 1985. 'Arrow-Odd.' In *Seven Viking Romances*. London: Penguin, pp. 25–137.
Hermann Pálsson and Anthony Faulkes, trans. 2012. *The Uppsala-Edda*. London: Viking Society for Northern Research, University College London.
Heusler, Andreas and Wilhelm Ranisch, eds. 1903. *Eddica minora: Dichtungen eddischer Art aus den Fornaldarsögur und anderen Prosawerken*. Dortmund: W. Ruhfus.
HGS = Magnus, Olaus. 1996. *Description of the Northern Peoples, 1555*, ed. P.G. Foote. Vols 1–3. London: Hakluyt Society.
Higden, Ranulf. 1865. *Polychronicon*, ed. Churchill Babington. London: Longman, Green, Longman, Roberts and Green.
Hversu Noregr byggðist. 1954. In Guðni Jónsson, ed. *Fornaldarsögur Norðurlanda*. Vol. 2. Reykjavík: Íslendingasagnaútgáfan, pp. 75–87.
ÍF 5 = Einar Ól. Sveinsson, ed. 1934. *Laxdœla saga*. Vol. 5. Íslenzk fornrit. Reykjavík: Hið íslenzka fornritafélag.
ÍF 6 = Guðni Jónsson, ed. 1943. *Vestfirðinga sögur: Gísla saga Súrssonar, Fóstbrœðra saga, Þáttr Þormóðar, Havarðar saga Ísfirðings, Auðunar þáttr vestfirzka, Þorvarðar þáttr Krákunefs*. Vol. 6. *Íslenzk fornrit*. Reykjavík: Hið íslenzka fornritafélag.
ÍF 8 = Einar Ól. Sveinsson, ed. 1958. *Vatnsdæla saga*. Vol. 8. Íslenzk fornrit. Reykjavík: Hið íslenzka fornritafélag.

ÍF 10 = Björn Sigfússon, ed. 1940. *Reykdœla saga*. Íslenzk Fornrit. Vol. 10. Reykjavík: Hið íslenzka fornritafélag.
ÍF 12 = Einar Ól. Sveinsson, ed. 1954. *Brennu-Njáls saga*. Vol. 12, Íslenzk Fornrit. Reykjavík: Hið íslenzka fornritafélag.
ÍF 13 = Þórhallur Vilmundarson and Bjarni Viljhálmsson, eds. 1991. *Harðar saga*. Vol. 13. Íslenzk Fornrit. Reykjavík: Hið íslenzka fornritafélag.
ÍF 14 = Jóhannes Halldórsson, ed. 2007. *Kjalnesinga saga*. Vol. 14. Íslenzk fornrit. Reykjavík: Hið íslenzka fornritafélag.
ÍF 15 = Guðrún Ása Grímsdóttir, ed. 2003. *Biskupa sögur I*. Vol. 15. *Íslenzk fornrit*. Reykjavík: Hið íslenzka fornritafélag.
ÍF 26 = Bjarni Aðalbjarnarson, ed. 2002. *Heimskringla I*. Vol. 26. *Íslenzk fornrit*. Reykjavík: Hið íslenzka fornritafélag.
ÍF 27 = Bjarni Aðalbjarnarson, ed. 2002. *Heimskringla II*. Vol. 27. Íslenzk fornrit. Reykjavík: Hið íslenzka fornritafélag.
ÍF 28 = Bjarni Aðalbjarnarson, ed. 2002. *Heimskringla III*. Vol. 28. *Íslenzk fornrit*. Reykjavík: Hið íslenzka fornritafélag.
JMSV = N. F. Blake, trans. 1962. *The Saga of the Jomsvikings*. London: Thomas Nelson and Sons.
Kålund, Kr., ed. 1889–1891. *Laxdæla saga*. Copenhagen: S. L. Møllers Bogtrykkeri (Møller & Thomsen).
Kramer, Heinrich. 2009. *The Hammer of Witches: A Complete Translation of the* Malleus Maleficarum. trans. Christopher S. Mackay. Cambridge: Cambridge University Press.
Kunin, Devra and Carl Phelpstead, trans. 2008. *A History of Norway, and the Passion and Miracles of Blessed Óláfr*. London: Viking Society for Northern Research, University College London.
Larrington, Carolyne. 2014. *The Poetic Edda*. Oxford: Oxford University Press.
Neckel, G., and H. Kuhn, eds. 1962. *Edda: die Lieder des Codex Regius nebst verwandten Denkmälern*. 4th ed. Heidelberg: Carl Winter.
NGL = Keyser, R., and P. A. Munch, eds. 1836–1849. *Norges gamle Love indtil 1387*. 3 vols. Christiania (Oslo): Gröndahl.
Ólafur Halldórsson, ed. 2000. Óláfs saga Tryggvasonar en mesta. 3 vols. *Editiones Arnamagnaeanae*, Series A. Copenhagen: Munksgaard.
Magnús Rafnsson, ed. 2008. *Tvær galdraskræður: Lbs 2413 8vo: Leyniletursskræðan Lbs 764 8vo*. Hólmavík: Strandagaldur.
Rimbert. 1884. *Vita Anskarii*, ed. Georg Waitz. Hannover: Hahnsche Buchhandlung.
Sandvik, H., and H. Winge, eds. 1987. *Tingbok for Finnmark, 1620–1633*. Oslo.
Saxo = Saxo Grammaticus. 2015. *Gesta Danorum*. trans. P. Fisher and ed. Karsten Friis-Jensen. 2 vols. Oxford: Clarendon Press.
SKALD 1 = Whaley, Diana, ed. 2012. Poetry from the Kings' Sagas 1: From Mythical Times to c. 1035. Skaldic Poetry of the Scandinavian Middle Ages 1. Turnhout: Brepols.
SKALD 2 = Gade, Kari Ellen, ed. 2009. Poetry from the Kings' Sagas 2: From c. 1035 to c. 1300. Skaldic Poetry of the Scandinavian Middle Ages 2. Turnhout: Brepols.
SKALD 8 = Clunies Ross, Margaret. 2017. Poetry in Fornaldarsögur. *Skaldic Poetry of the Scandinavian Middle Ages 8*. Turnhout: Brepols.
Slay, D., ed. 1960. 'Hrolfs Saga Kraka.' In *Editiones Arnamagnæanæ B*. Copenhagen: Munksgaard.
Van Arsdall, A. 2002. *Medieval Herbal Remedies: The Old English Herbarium and Anglo-Saxon Medicine*. London: Routledge.
Von Herberstein, Sigmund Freiherr, trans. 1851. *Notes Upon Russia: Being a Translation of the Earliest Account of that Country, Entitled Rerum Moscoviticarum Commentarii*. London: Hakluyt Society.

Waggoner, Ben, trans. 2012. *The Hrafnista Sagas*. New Haven: Troth Publications.
Waggoner, Ben, trans. 2014. *Six Sagas of Adventure*. New Haven: Troth Publications.
Waggoner, Ben, trans. 2017. *Sagas of Imagination: A Medieval Icelandic Reader*. Philadelphia: Troth Publications.
Þorláks = Ármann Jakobsson and David Clark, eds. 2013. *Þorláks saga byskups*. London: Viking Society for Northern Research, University College London.

Secondary Sources

Aalto, Sirpa. 2003. 'Alienness in *Heimskringla*: Special Emphasis on the Finnar.' *Scandinavia and Christian Europe in the Middle Ages: Papers of the 12th International Saga Conference*, ed. Rudolf Simek and Judith Meurer. Bonn, Germany. Bonn: Universität Bonn, pp. 1–7.
Aalto, Sirpa. 2010. 'Categorizing Otherness in the King's Sagas.' PhD thesis, University of Eastern Finland.
Aðalheiður Guðmundsdóttir. 2015. 'The Narrative Role of Magic in the *Fornaldarsögur*.' *ARV: Nordic Yearbook of Folklore* 70: 39–56.
Alm, Ellen, and Rune Blix Hagen. 2020. 'Sámi Magic and Rituals from *Historia Norwegie* to Johannes Schefferus, c. 1150–1680.' In *What is North? Imagining the North from Ancient Times to the Present Day*, ed. Oisín Plumb, Alexandra Sanmark, Donna Heddle. Turnhout: Brepols, pp. 153–173.
Anderson, Alan Orr. 1922. *Early Sources of Scottish History, AD 500 to 1286*. 2 vols. Edinburgh and London: Oliver and Boyd.
Andrén, Anders. 2014. *Tracing Old Norse Cosmology: the World Tree, Middle Earth, and the Sun from Archaeological Perspectives*. Lund: Nordic Academic Press.
Arrhenius, Birgit. 2011. 'Helgö – Pagan Sanctuary Complex.' In *Excavations at Helgö XVIII. Conclusions and New Aspects*, ed. Birgit Arrhenius and U. O'Meadhra. Stockholm: Kungliga Vitterhets Historie och Antikvitets Akademien, pp. 11–43.
Ármann, Jakobsson. 2017. *The Troll Inside You: Paranormal Activity in the Medieval North*. Santa Barbara, California: Punctum Books.
Bäck, M., A. M. H. Stenholm, and J. Å. Ljun. 2008. *Lilla Ullevi – historien om det fridlysta rummet: Vendeltida helgedom, medeltida by och 1600-talsgård: Uppland, Bro socken, Klöv och Lilla Ullevi 1: 5, Jursta 3: 3, RAÄ 145: arkeologisk undersökning*. Stockholm: Riksantikvarieämbetets arkeologiska uppdragsverksamhet.
Barraclough, Eleanor Rosamund. 2017. 'Arctic Frontiers: Rethinking Norse–Sámi Relations in the Old Norse Sagas.' In *Medieval and Renaissance Studies* 48 (3): 27–51.
Barraclough, Eleanor Rosamund. 2020. 'The Ice Giant Cometh: The Far North in the Old Norse–Icelandic Sagas.' In *Myths and Magic in the Medieval Far North: Realities and Representations of a Region on the Edge of Europe*, ed. Stefan Figenschow, Richard Holt and Miriam Tveit. Turnhout: Brepols, pp. 71–94.
Belier, Wouter W. 1991. *Decayed Gods: Origin and Development of Georges Dumézil's Idéologie Tripartite*. Leiden: Brill.
Berner, Ulrich. 2001. 'The Notion of Syncretism in Historical and/or Empirical Research.' *Historical Reflections/Reflexions Historiques* 27 (3): 499–509.
Bertell, Maths. 2003. *Tor och den nordiska åskan: Föreställningar kring världsaxeln*. Stockholm: Religionshistoriska institutionen, Stockholms universitet.
Birkett, Thomas. 2016. *Reading the Runes in Old English and Old Norse Poetry*. Abingdon-on-Thames: Taylor & Francis.
Boyer, Régis. 1973. 'The Influence of Pope Gregory's Dialogues on Old Icelandic Literature.' In *Proceedings of the First International Saga Conference, University of Edinburgh, 1971*, ed. Peter Foote, Hermann Pálsson and Desmond Slay. London: Viking Society for Northern Research, University College London.

Burke, Peter. 1992. *History and Social History*. Ithaca: Cornell University Press.
Burrows, Hannah. 2021. 'Showered with Praise: Weatherscapes in Late Tenth-century Skaldic Verse.' *Medieval Ecocriticisms* 1 (1): 65–87.
Byock, Jesse L., trans. 1990. *The Saga of the Volsungs*. London: Penguin Classics.
Cameron, Euan. 2010. *Enchanted Europe: Superstition, Reason, and Religion 1250–1750*. Oxford: Oxford University Press.
Cleasby, Richard, and Guðbrandur Vigfússon. 1874. *An Icelandic–English Dictionary*. Oxford: Clarendon Press.
Conroy, Patricia. 1980. '*Laxdœla saga* and *Eiríks saga rauða*: Narrative Structure.' *Arkiv för nordisk filologi* 95: 116–125.
Damm, Charlotte. 2005. 'Archaeology, Ethno-History and Oral Traditions: Approaches to the Indigenous Past.' In *Norwegian Archaeological Review* 38 (2): 73–87.
Day, Cyrus Lawrence. 1967. *Quipus and Witches' Knots: The Role of the Knot in Primitive and Ancient Culture, with a Translation and Analysis of 'Oribasius de Laqueis.'* Lawrence, Kansas: University Press of Kansas.
de Vries, Jan. 1970. *Altgermanische Religionsgeschichte*. Berlin: De Gruyter & Co.
de Vries, Jan. 1977. *Altnordisches etymologisches Wörterbuch*. Vol. 2. Leiden: Brill Archive.
DeAngelo, Jeremy. 2010. 'The North and the Depiction of the "Finnar" in the Icelandic Sagas.' *Scandinavian Studies* 82 (3): 257–286.
DeAngelo, Jeremy. 2018. 'Making Sami of the Scots: Britain's and Scandinavia's Near Norths.' In *Visions of North in Premodern Europe*, ed. D. Jørgensen and V. Langum. Turnhout: Brepols, pp. 207–221.
Dillmann, François-Xavier. 2006. *Les Magiciens dans l'Islande ancienne: Études sur la représentation de la magie Islandaise et de ses agents dans les sources littéraires norroises*. Uppsala: Gustav Adolfs Akademien för Svensk Folkkultur.
Dillmann, François-Xavier. 2007. 'Zauber.' *Reallexikon der germanischen Altertumskunde* 35: 58–86.
DuBois, Thomas Andrew. 1999. *Nordic Religions in the Viking Age*. Philadelphia: University of Pennsylvania Press.
Dumézil, Georges. 1973. *Gods of the Ancient Northmen*. trans. Einar Haugen. Vol. 3. Berkeley: University of California Press.
Dybdahl, Audun. 2012. 'Climate and Demographic Crises in Norway in Medieval and Early Modern Times.' *The Holocene* 22 (10): 1159–1167.
Einar Ól Sveinsson. 1971. 'Svipdag's Long Journey: Some Observations on Grógaldr and Fjölsvinnsmál.' In *Hereditas: Essays and Studies presented to Professor Séamus Ó Duilearga, former Honorary Director of the Irish Folklore Commission*, ed. Bo Almquist, Breandán Mac Aodha, and Gearóid Mac Eoin. Dublin: Folklore of Ireland Society, pp. 298–319.
Falk, Hjalmar. 1924. *Odensheite*. Skrifter utg. av Videnskapsselskapet i Kristiania. II. Historisk-filosofisk Klasse. 1924: 10. Kristiania (Oslo): Dybwad.
Figenschow, Stefan, Richard Holt and Miriam Tveit, eds. 2021. *Myths and Magic in the Medieval Far North: Realities and Representations of a Region on the Edge of Europe*. Turnhout: Brepols.
Flint, Valerie I. J. 1991. *The Rise of Magic in Early Medieval Europe*. Princeton: Princeton University Press.
Flint, Valerie I. J. 2002. 'Magic in English Thirteenth-Century Miracle Collections.' In *The Metamorphosis of Magic from Late Antiquity to the Early Modern Period*, ed. Jan N. Bremmer, Jan R. Veenstra and Brannon Wheeler. Leuven: Peeters.
Foran, Derek. 1995. 'The Three Wind-Knots: A Maritime Migratory Legend in Irish Tradition.' *Folklore* 8: 56–59.
Gardeła, Leszek, Sophie Bønding and Peter Pentz, eds. 2023. *The Norse Sorceress: Mind and Materiality in the Viking World*. Oxford: Oxbow Books.
Gerhardt, Walter. 1966. 'The Hebrew/Israelite Weather-Deity.' *Numen* 13 (2): 128–143.

Gísli Pálsson. 1991. *The Name of the Witch: Sagas, Sorcery and Social Content*. Glasgow: Cruithne Press.

Glauser, Jürg and Pernille Hermann, eds. 2021. *Myth, Magic, and Early Memory in Early Scandinavian Narrative Culture: Studies in Honour of Stephen A. Mitchell*. Turnhout: Brepols.

Goodich, Michael E. 2007. *Miracles and Wonders: The Development of the Concept of Miracle, 1150–1350*. Aldershot: Ashgate.

Graham-Campbell, J. 1980. *Viking Artefacts: A Select Catalogue*. London: British Museum Publications.

Graham-Campbell, James. 2003. 'Review of *Thor the Wind-Raiser and the Eyrarland Image*.' *Speculum* 78 (4): 1370–1371.

Gräslund, Bo. 2007. 'Fimbulvintern, Ragnarök och klimatkrisen år 536–537 e. Kr.' *Saga och sed: Kungl. Gustav Adolfs akademiens årsbok*: 93–123.

Grønvik, Ottar. 1992. 'En hedensk bønn. Runeinnskriften på en liten kopperplate fra Kvinneby på Öland.' In *Eyvindarbók. Festskrift til Eyvind Fjeld Halvorsen*, ed. Finn Hødnebø et al.. Oslo: Institutt for nordistikk og litteraturvitenskap, pp. 71–85.

Gunnell, Terry. 2014. '"Magical Mooning" and the "Goatskin Twirl": "Other" Kinds of Female Magical Practices in Early Iceland.' In *Nordic Mythologies: Interpretations, Intersections, and Institutions*. Berkeley, California: North Pinehurst Press, pp. 133–153.

Gunnell, Terry. 2015. 'Pantheon? What Pantheon? Concepts of a Family of Gods in Pre-Christian Scandinavian Religions.' *Scripta Islandica: Isländska Sällskapets Årsbok* 66: 55–76.

Gunnell, Terry. 2017. 'How High Was the High One? The Roles of Óðinn and Þórr in Pre-Christian Icelandic Society.' In Theorizing Old Norse Myth, ed. Stefan Brink and Lisa Collinson. Turnhout: Brepols, pp. 105–129.

Gustavson, Helmer. 2014. 'Sigtunas lösföremål.' Accessed 14 June 2024. http://www.raa.se/kulturarvet/arkeologi-fornlamningar-och-fynd/runstenar/digitala-sveriges-runinskrifter/sigtunas-losforemal/.

Hagen, Rune Blix. 1999. 'The Witch-Hunt in Early Modern Finnmark.' *Acta Borealia* 16 (1): 43–62.

Hagen, Rune Blix. 2002. 'Harmløs dissenter eller djevelsk trollmann?' *Historisk tidsskrift* 81 (2–3): 319–346.

Hagen, Rune Blix. 2014. 'Witchcraft and Ethnicity: A Critical Perspective on Sami Shamanism in Seventeenth-Century Northern Norway.' In *Writing Witch-Hunt Histories: Challenging the Paradigm*, ed. Marko Nenonen and Raisa Maria Toivo. Boston: Brill, pp. 141–166.

Hagen, Rune Blix. 2017. 'Images, Representations and the Self-Perception of Magic among the Sami Shamans of Arctic Norway, 1592–1692.' In *Contesting Orthodoxy in Medieval and Early Modern Europe: Heresy, Magic and Witchcraft*, ed. Louise Nyholm Kallestrup, Raisa Maria Toivo. London: Palgrave Macmillan, pp. 279–300.

Halinen, Petri. 2016. 'The Inland Sámi Societies of Northern Fennoscandia during the Late Iron Age and Early Medieval Period: An Archaeological Approach.' *Iskos* 21: 160–174.

Halvorsen, Eyvind Fjeld. 1960. 'Galder.' *Kulturhistorisk leksikon for nordisk middelalder* 5: 159–161.

handrit.is. 'handrit.is.' Accessed 14 June 2024. https://handrit.is/en.

Hansen, Lars Ivar, and Bjørnar Olsen. 2014. *Hunters in Transition: An Outline of Early Sámi History*. Leiden: Brill.

Harlan-Haughey, Sarah. 2015. *A Landscape of Conflict: Three Stories of the Faroe Conversions*. Ithaca, New York: Cornell University Library.

Harris, Joseph and Thomas Hill. 1989. 'Gestr's "Prime Sign": Source and Signification in *Norna-Gests þáttr*.' *Arkiv för nordisk filologi* 104: 103–122.

Harris, Joseph. 2008. 'Saga as Historical Novel.' In *Structure and Meaning in Old Norse Literature: New Approaches to Textual Analysis and Literary Criticism*, ed. John Lindow, Lars Lönnroth, and Gerd Wolfgang Weber. Odense: Odense University Press, pp. 228–260.

Hastrup, Kirsten. 1990. *Island of Anthropology: Studies in Past and Present Iceland*. Odense: Odense University Press.

Heide, Eldar. 2006. 'Spinning *Seiðr*.' In *Old Norse Religion in Long-Term Perspectives: Origins, Changes and Interactions*, ed. Anders Andrén, Kristina Jennbert and Catharina Raudvere. Lund: Nordic Academic Press, pp. 164–168.

Heide, Eldar. 2011. 'Holy Islands and the Otherworld: Places Beyond Water.' *Isolated Islands in Medieval Nature, Culture and Mind*, ed. Gerhard Jaritz and Torsten Jørgensen. Berlin: De Gruyter, pp. 57–80.

Hemphill, Jennifer. 2020. 'Review: Neil Price. *The Viking Way: Magic and Mind in Late Iron Age Scandinavia*.' *Kyngervi* 1: 93–97.

Hemphill, Jennifer, Ségdae Richardson-Read and Solveig Marie Wang, eds. 2024. *Performing Magic in the Pre-Modern North*. London: Palgrave MacMillan.

Hermann Pálsson. 1997. *Úr landnorðri: Samar og ystu rætur íslenskrar menningar*. Studia Islandica, 54. Reykjavík: Bókmenntafræðistofnun Háskóla Íslands.

International Labour Organization. 1989. 'ILO 169: Indigenous and Tribal Peoples Convention, 1989 (NO. 169).' Accessed 28 April 2024. https://www.ohchr.org/en/instruments-mechanisms/instruments/indigenous-and-tribal-peoples-convention-1989-no-169.

Jesch, Judith. 2008. 'Poetry in the Viking Age.' In *The Viking World*, ed. Stefan Brink and Neil Price. London: Routledge, pp. 315–327.

Jochens, Jenny. 1996. *Old Norse Images of Women*. Pennsylvania: University of Pennsylvania Press.

Jóhanna Katrín Friðriksdóttir. 2009. 'Women's Weapons: A Re-Evaluation of Magic in the *Íslendingasögur*.' *Scandinavian Studies* 81 (4): 409–436.

Jóhanna Katrín Friðriksdóttir. 2013. *Women in Old Norse Literature: Bodies, Words, and Power*. London: Palgrave Macmillan.

Jóhanna Katrín Friðriksdóttir. 2021. *Valkyrie: The Women of the Viking World*. London: Bloomsbury Academic.

Johnston, George, trans. 1994. *Thrand of Gotu: Two Icelandic Sagas from the Flat Island Book*. Erin, Ontario: Porcupine's Quill Press.

Jonsson, Bengt R., Svale Solheim and Eva Danielson, eds. 1978. *The Types of the Scandinavian Medieval Ballad: A Descriptive Catalogue*. Odense: Universitetsforlaget.

Kaikkonen, Konsta Ilari. 2020a. *Contextualising Descriptions of Noaidevuohta: Saami Ritual Specialists in Text Written in 1871*. Unpublished doctoral thesis, University of Bergen.

Kaikkonen, Konsta Ilari. 2020b. 'Sámi Indigenous(?) Religion(s)(?)—Some Observations and Suggestions Concerning Term Use.' *Religions* 11 (9): 1–14.

Kalinke, Marianne. 1985. 'Norse Romance (Riddarasögur).' In *Old Norse–Icelandic Literature: A Critical Guide*, ed. Carol J. Clover and John Lindow. Toronto: University of Toronto Press, pp. 316–363.

Kallestrup, Louise Nyholm. 2025. *The Construction of Witchcraft in Early Modern Denmark, 1536–1617*. London: Routledge.

Karlsson, Stefán. 1970. 'Um Vatnshyrnu.' *Opuscula* 4: 279–303.

Kershaw, Priscilla Kasden. 2000. *The One-Eyed God: Odin and the (Indo-)Germanic Männerbünde*. Washington, DC: Institute for the Study of Man.

Kick, Donata, John D. Shafer, and John McKinnell, eds. 2014. *Essays on Eddic Poetry*. Toronto: University of Toronto Press.

Kjærum, Poul, and Rikke Agnete Olsen. 1990. 'Oldtidens Ansigt: Faces of the Past.' In *Hendes Majestæt Dronning Margrethe II*, ed. Poul Kjærum and Rikke Agnete Olsen. Copenhagen: Det kongelige nordiske oldskriftselskab.

Korecká, Lucie. 2017. '*Óðin stoyttist í jörðina niður*: Magic and Myth in the Faroese Ballads.' *ARV Nordic Yearbook of Folklore* 73: 91–114.

Koskela Vasaru, Mervi. 2012. 'Bjarmaland and Interaction in the North of Europe from the Viking Age until the Early Middle Ages.' *Journal of Northern Studies* 6 (2): 37–58.

Králová, Kristýna. 2017. 'What Did the Future Hold for them? Different Types of Foreshadowing in Various Saga Genres.' *European Journal of Scandinavian Studies* 47 (1): 24–50.

Kulturens, Årsbok. 1838. *Kulturens: en årsbok till medlemmarna av kulturhistoriksa föreningen för södra Sverige*. Lund: Kulturen i Lund.

Kusmenko, Jurij K. 2009. 'Sámi and Scandinavians in the Viking Age.' *Scandinavistica Vilnensis* (2): 65–94.

Källström, Magnus. 2010. 'Det 22: a internationella fältrunologmötet 11–13 september 2009 i Sigtuna.' *Fornvännen* 105 (2): 121–123.

Langeslag, Paul S. 2015. *Seasons in the Literatures of the Medieval North*. Woodbridge: Boydell & Brewer.

Larrington, Carolyne. 1993. *A Store of Common Sense: Gnomic Theme and Wisdom in Old Icelandic and Old English Wisdom Poetry*. Oxford: Clarendon Press.

Lassen, Annette. 2021. *Odin's Ways: A Guide to Odin in the Medieval World*. Abingdon-on-Thames: Taylor & Francis.

Lavender, Philip. 2018. '"*Sumar eptir fornkvæðum eðr fróðum mönnum ok stundum eptir fornum bókum*": Some Observations on the Sources of *Göngu-Hrólfs saga*.' *Scandinavian Studies* 90 (1): 78–109.

Leslie, Helen F. 2009. 'Border Crossings: Landscape and the Other World in the *Fornaldarsögur*.' *Scripta Islandica* 60: 119–135.

Liestøl, Aslak et al. 1962. *Drottkvæt-vers fra Bryggen i Bergen*, *Maal og Minne*. Sofiemyr: Novus forlag.

Lindow, John. 1995. 'Supernatural Others and Ethnic Others: A Millennium of World View.' *Scandinavian Studies* 67 (1): 8–31.

Lindquist, Ivar. 1963. 'Två vikingatida gudabeläten.' *Kulturen 1962*: 70–78.

Linquist, Ivar. 1987. *Religiösa runtexter III. Kvinneby-amuletten: ett tydningsförslag efter författarens efterlämnade manuskript utg. av Gösta Holm*. Lund.

Lionarons, Joyce T. 1997. 'Gender, Ethnicity, and Pagan Magic in Snorri Sturluson's *Heimskringla*.' *Sagas and the Norwegian Experience. 10th International Saga Conference. Trondheim, 3.–9. August 1997*. Preprints. Trondheim: Senter for Middelalderstudier, pp. 419–426.

Livesay, Jennifer. 1988. 'Women and Narrative Structure in "*Eyrbyggia Saga*".' *Folklore Forum* 21 (2): 181–193.

Livingston, Michael. 2004. 'More Vinland Maps and Texts Discovering the New World in Higden's *Polychronicon*.' *Journal of Medieval History* 30 (1): 25–44.

Loomis, Charles Grant. 1948. *White Magic: An Introduction to the Folklore of Christian Legend*. Cambridge, Massachusetts: The Mediaeval Academy of America.

Louis-Jensen, Jonna. 2001. 'Halt illu frān Būfa! Til tolkningen af Kvinneby-amuletten fra Öland.' In *Northern Lights: Following Folklore in North Western Europe. Essays in Honor of Bo Almqvist*, ed. Séamas Ó Catháin. Dublin: University College Dublin, pp. 111–126.

Louis-Jensen, Jonna. 2005. 'Södra Kvinneby.' *Reallexikon der germanischen Altertumskunde*, ed. Johannes Hoops and Heinrich Beck. Berlin: De Gruyter, pp. 193–196.

MacLeod, Mindy, and Bernard Mees. 2006. *Runic Amulets and Magic Objects*. Woodbridge: Boydell Press.

Magnusson, Magnus. 1976. *Hammer of the North*. London: Orbis.

McCartney, Eugene S. 1925. 'Magic and the Weather in Classical Antiquity.' *The Classical Weekly* 18: 154–157.

McCreesh, Bernadine. 2006. 'Elements of the Pagan Supernatural in the Bishops' Sagas.' In *The Fantastic in Old Norse Icelandic Literature: Preprint Papers of the 13th International Saga Conference, Durham and York, 6th–12th August 2006*, ed. John McKinnell, David Ashurst, and Donata Kick. Durham: Centre for Medieval and Renaissance Studies, pp. 671–680.

McCreesh, Bernadine. 2018. *The Weather in the Icelandic Sagas: The Enemy Without*. Cambridge: Cambridge Scholars Publishing.

McKinnell, John. 1970. 'The Reconstruction of Pseudo-Vatnshyrna.' *Opuscula* 4: 304–338.

McKinnell, John. 2002. 'Þorgerðr Hǫlgabrúðr and Hyndluljóð'. In *Mythological Women: Studies in Memory of Lotte Motz (1922–1997)*, ed. Rudolf Simek and Wilhelm Heizmann. Wien: Fassbaender, pp. 265–290.

McKinnell, John, Rudolf Simek, and Klaus Düwel, eds. 2004. *Runes, Magic and Religion. A Sourcebook*. Vienna: Fassbaender.

McKinnell, John. 2005. *Meeting the Other in Norse Myth and Legend*. Cardiff: D. S. Brewer.

McKinnell, John. 2007. 'The Making of *Hávamál*.' *Viking and Medieval Scandinavia* 3: 75–115.

Meissner, Rudolf. 1921. *Die Kenningar der Skalden: Ein Beitrag zur skaldischen Poetik*. Vol. 1. Bonn and Leipzig: Schroeder.

Merkelbach, Rebecca. 2019. *Monsters in Society: Alterity, Transgression, and the Use of the Past in Medieval Iceland*. Berlin: DeGruyter and Medieval Institute Publications.

Meylan, Nicolas. 2011. 'Magic and Discourses of Magic in the Old Norse Sagas of the Apostles.' *Viking and Medieval Scandinavia* 7: 107–124.

Meylan, Nicolas. 2014. *Magic and Kingship in Medieval Iceland: The Construction of a Discourse of Political Resistance*. Turnhout: Brepols.

Miller, William Ian. 1986. 'Dreams, Prophecy and Sorcery: Blaming the Secret Offender in Medieval Iceland.' *Scandinavian Studies* 58 (2): 101–123.

Mitchell, Stephen. 1985. 'Scandinavian Balladry in the Old Norse Legacy: "Álvur kongur" (CCF 14) "Stolt Herr Alf" (ST 5), and Hálfs saga.' *ARV: The Yearbook of Scandinavian Folklore* 41: 1–11.

Mitchell, Stephen. 1991. Heroic Sagas and Ballads. Ithaca, NY: Cornell University Press.

Mitchell, Stephen A. 2001. 'Warlocks, Valkyries and Varlets: A Prolegomenon to the Study of North Sea Witchcraft Terminology.' *Cosmos: The Journal of the Traditional Cosmology Society* 17 (1): 59–81.

Mitchell, Stephen A. 2011. *Witchcraft and Magic in the Nordic Middle Ages*. Pennsylvania: University of Pennsylvania Press.

Mitchell, Stephen. 2013. 'Magic as Acquired Art and the Ethnographic Value of the Sagas.' In *Old Norse Myths, Literature and Society*, ed. Margaret Clunies Ross. Viborg: University Press of Southern Denmark, pp. 132–152.

Mitchell, Stephen. 2015. 'Leechbooks, Manuals, and Grimoires: On the Early History of Magical Texts in Scandinavia.' *ARV: Nordic Yearbook of Folklore* 70: 57–74.

Mitchell, Stephen A. 2019. 'Scandinavia.' In *The Routledge History of Medieval Magic*, ed. Sophie Page and Catherine Rider. London and New York: Routledge, pp. 136–150.

Mormando, Franco. 1999. *The Preacher's Demons: Bernardino of Siena and the Social Underworld of Early Renaissance Italy*. Chicago: University of Chicago Press.

Morris, Katherine. 1991. *Sorceress or Witch?: The Image of Gender in Medieval Iceland and Northern Europe*. Maryland: University Press of America.

Moyne, Ernest J. 1981. *Raising the Wind: The Legend of Lapland and Finland Wizards in Literature*. Newark: University of Delaware Press.

Mundal, Else. 1996. 'The Perception of the Saamis and their Religion in Old Norse Sources.' *Shamanism and Northern Ecology* 36: 97–116.

Mundal, Else. 2018. 'Sami Sieidis in a Nordic Context?' *Journal of Northern Studies* 12 (1): 11–20.
Niemi, Einar. 1993. 'Hans Hansen Lilienskiold–embetsmann, vitenskapsmann og opprører.' In *Portretter fra norsk historie*. Oslo: Universitetsforlaget, pp. 43–72.
Nilsson, Bruce E. 1976. 'The Runic "Fish-Amulet" from Öland: A Solution.' *Mediaeval Scandinavia* 9: 236–245.
Nordén, Arthur. 1943. 'Bidrag till svensk runforskning.' *ANF* 53: 147–189.
O'Connor, Ralph. 2009. 'Truth and Lies in the *fornaldarsögur*: The Prologue to Göngu-Hrolfs saga.' In *Fornaldarsagaerne: myter og virkelighed*, ed. Ármann Jakobsen, Agneta Ney, and Annette Lassen. Copenhagen: Museum Tusculanum, University of Copenhagen, pp. 361–370.
Ogilvie, Astrid E. J., and Gísli Pálsson. 2003. 'Mood, Magic, and Metaphor: Allusions to Weather and Climate in the Sagas of Icelanders.' In *Weather, Climate, Culture*, ed. Sarah Strauss and Benjamin Orlove. London: Routledge, pp. 251–274.
Ojala, Carl-Gösta. 2009. *Sámi Prehistories: The Politics of Archaeology and Identity in Northernmost Europe*. Uppsala: Institutionen för arkeologi och antik historia.
Otto, Bernd-Christian, and Michael Stausberg, eds. 2013. *Defining Magic: A Reader*. London: Routledge.
Page, R. I. 1962–1965. 'Lapland Sorcerers.' *Saga Book* 16: 215–232.
Perabo, Lyonel D. 2016. *Here be Heathens: The Supernatural Image of Northern Fenno-Scandinavia in Pre-Modern Literature*. Unpublished Master's Dissertation, University of Iceland.
Pereswetoff-Morath, Sofia. 2017. *Vikingatida runbleck: Läsningar och tolkningar*. Uppsala: Institutionen för nordiska språk, Uppsala universitet.
Perkins, Richard. 1972. 'An Edition of *Flóamanna saga* with a Study of its Sources and Analogues.' Unpublished doctoral thesis, University of Oxford.
Perkins, Richard. 2001. *Thor the Wind-raiser and the Eyrarland Image*. London: Viking Society for Northern Research, University College London.
Phelpstead, Carl. 2020. *An Introduction to the Sagas of Icelanders*. Gainesville: University Press of Florida.
Porsanger, Jelena. 2018. Sámi eamioskkoldaga doahpagastin. *Dîn: Tidsskrift for Religion og Kultur* 2018: 146–157.
Price, Neil. 2002. *The Viking Way: Religion and War in Late Iron Age Scandinavia*. Uppsala: Department of Archaeology and Ancient History.
Price, Neil. 2019. *The Viking Way: Magic and Mind in Late Iron Age Scandinavia*. Oxford: Oxbow Books.
Rafnsson, Sveinbjörn. 2017. 'Landnámabók and Its Sturlubók Version.' In *Sturla Þórðarson: Skald, Chieftain and Lawman*, ed. Jón Viðar Sigurðsson and Sverrir Jakobsson. Leiden: Brill, pp. 44–55.
Raudvere, Catharina. 1993. *Föreställningar om maran i nordisk folktro*. Vol. 1. Lund: Religionshistoriska avd.: Lunds universitet.
Raudvere, Catharina. 2002. '*Trolldómr* in Early Medieval Scandinavia.' In *Witchcraft and Magic in Europe*, Volume 3: *The Middle Ages*, ed. Bengt Ankarloo and Stuart Clark. Philadelphia, Pennsylvania: University of Pennsylvania Press, pp. 73–171.
Raudvere, Catharina. 2003. *Kunskap och insikt i norrön tradition: mytologi, ritualer och trolldomsanklagelser*. Lund: Nordic Academic Press.
Raudvere, Catharina. 2005. 'The Power of the Spoken Word as Literary Motif and Ritual Practice in Old Norse Literature.' *Viking and Medieval Scandinavia* 1: 179–202.
Robbins, Rossell Hope. 2015. *The Encyclopedia of Witchcraft and Demonology*. Brattleboro, Vermont: Echo Point Books & Media, LLC.
Roesdahl, Else, ed. 1992. *Viking og Hvidekrist: Norden og Europa 800–1200*. Den 22. Europarådsudstilling. Copenhagen: Nationalmuseet.
Ross, Margaret Clunies. 2005. *A History of Old Norse Poetry and Poetics*. Cambridge: D. S. Brewer.

Saarikivi, Janne. 2011. 'Saamelaiskielet-nykypäivää ja historiaa.' *Saamentutkimus tänään* 234: 177–219.

Schjødt, Jens Peter. 2008. 'The Old Norse Gods.' In *The Viking World*, ed. Stefan Brink and Niel Price. London: Routledge, pp. 219–222.

Schjødt, Jens Peter. 2019. 'Mercury–Wotan–Óðinn: One or Many?' In *Myth, Materiality, and Lived Religion in Merovingian and Viking Scandinavia*, ed. Klas Wikström af Edholm, Peter Jackson Rova, Andreas Nordberg, Olof Sundqvist and Torun Zachrisson. Stockholm: Stockholm University Press, pp. 59–86.

Simek, Rudolf. 1993. *Dictionary of Northern Mythology*, trans. Angela Hall. Cambridge: Brewer.

Solli, Brit. 2002. *Seid: myter, sjamanisme og kjønn i vikingenes tid*. Oslo: Pax Forlag A/S.

Sørensen, Preben Meulengracht and Kirsten Williams. 2002. '"Þorr's Fishing Expedition" [*Hymiskviða*].' In *The Poetic Edda: Essays on Old Norse Mythology*, ed. Paul Acker and Carolyne Larrington. London: Routledge, pp. 119–138.

Stephens, George (1878) *Thunor the Thunderer, Carved on a Scandinavian Font About the Year 1000*. London: Williams and Norgate, p. 33.

Strömbäck, Dag. 1935. *Sejd: Textstudier i nordisk religionshistoria*. Stockholm: H. Geber.

Strömbäck, Dag. 2000. *Sejd: och andra studier i nordisk själsuppfattning*. Stockholm: Kungl. Gustav Adolfs Akademien för svensk folkkultur, Gidlunds förlag.

Sundqvist, Olof. 2012. 'Var sejdhällen (fvn. seiðhjallr, hjallr) en permanent konstruktion vid kultplatser och kultbyggnader.' *Fornvännen* 107 (4): 280–285.

Svärdström, Elisabeth. 1967. 'Högstenableckets rungalder.' *Fornvännen* 62: 12–21.

Taggart, Declan. 2017. *How Thor Lost His Thunder: The Changing Faces of an Old Norse God*. London: Routledge.

Tolley, Clive. 1995. '*Vǫrðr* and *Gandr*: Helping Spirits in Norse Magic.' *Arkiv för nordisk filologi* 110: 56–75.

Tolley, Clive. 2009. *Shamanism in Norse Myth and Magic*. Helsinki: Academia Scientiarum Fennica.

Trynkina, Daria A. 2020. 'The Sámi People as Witches and Fairies in the British Tradition.' *Traditional Culture* 21 (1): 103–113.

Vaitkevičienė, Daiva. 2011. 'Libation in Baltic Religious Practices.' *Archaeologia Baltica* 15: 105–119.

Wang, Solveig Marie. 2023. *Decolonising Medieval Fennoscandia: An Interdisciplinary Study of Norse–Saami Relations in the Medieval Period*. Berlin: De Gruyter.

Wang, Solveig Marie. 2024. 'The Intersection of Medieval Studies and Indigenous Studies: A Norse–Saami Case Study.' *Postmedieval* 18: 119–150.

Westcoat, Eirik. 2016. 'The Goals of *Galdralag*: Identifying the Historical Instances and Uses of the Metre.' *Saga-Book* 40: 69–90.

Westlund, Börje. 1989. 'Kvinneby–en runinskrift med hittills okända gudanamn?' *Studia anthroponymica Scandinavica* 7: 25–52.

Wikman, K. R. V. 1917. 'Tabu- och orenhetsbegrepp i nordgermansk folktro om könen.' *Folkloristiska och etnografiska studier* D2: 1–62.

Willumsen, Liv Helene. 2013. *Witches of the North: Scotland and Finnmark*. Leiden: Brill.

Willumsen, Liv Helene. 2020. 'Anders Poulsen—Sámi Shaman Accused of Witchcraft, 1692.' *Folklore* 131 (2): 135–158.

Äikäs, Tiina, and Anna-Kaisa Salmi. 2013. '"The Sieidi is a Better Altar/the Noaidi Drum's a Purer Church Bell": Long-Term Changes and Syncretism at Sámi Offering Sites.' *World Archaeology* 45 (1): 64–82.

INDEX

For Product Safety Concerns and Information please contact our EU representative GPSR@taylorandfrancis.com Taylor & Francis Verlag GmbH, Kaufingerstraße 24, 80331 München, Germany

Batch number: 10725966

Printed by Printforce, the Netherlands